Sergio Rodriguez

The Scattering of Light by Matter

APPUNTI

SCUOLA NORMALE SUPERIORE
2001

ISBN: 978-88-7642-298-0

For Katinka, Katrin, Cecilia, Diana, and Kenneth
and
In memory of Professors Luciano Claude, Carlos Mori
and Domingo Almendras.

Preface

These notes contain my lectures on light scattering by matter presented at Scuola Normal Superiore, Pisa, Italy during May and June 1995. I have deleted some of the topics discussed then and added a few related to my recent work. The notes are not to be regarded as exhaustive but rather as a selection of topics. In particular I have discussed, as examples, the theoretical basis for the interpretation of experiment on light scattering by phonons in α-quartz and by electronic excitations in boron-doped diamond.

I wish to thank Professors G. Franco Bassani and Giuseppe C. La Rocca for their kind invitation to present these topics at Scuola Normale Superiore and to Professor A. K. Ramdas, Dr. Z. Barticevic, Dr. V. J. Tekippe, Dr. M. Grimsditch, Dr. H. Kim, and Dr. T. R. Anthony for their collaboration.

I am grateful to Mrs. Virginia Messick for her accurate, efficient, and cheerful preparation of the manuscript.

Contents

Chapter 1

The Classical Theory of the Scattering of Radiation

While observing the passage of light through a transparent medium we notice that there are not only reflected and refracted rays in addition to the incident beam but also a weak radiation propagating in all directions. This is the scattered radiation. Its presence can be explained by classical electromagnetic theory as follows. Consider, for simplicity, a gas of molecules possessing no permanent electric dipole moments. The action of the electric field of the light wave induces, in each molecule, an oscillatory electric dipole moment having the frequency of the incident wave. This oscillating dipole moment becomes the source of a secondary wave. The scattered light is the result of the secondary waves emitted by all molecules of the system.

To develop these concepts further we consider a single molecule with oscillating dipole moment

$$\boldsymbol{d}(t) = \Re\left(\boldsymbol{d}_0\, e^{-i\omega t}\right) = \boldsymbol{d}_0 \cos\omega t \,, \tag{1.1}$$

where we take the vector $\boldsymbol{d}_0$ real. This can always be done by selecting the origin of time in such a manner that the initial phase of the oscillator be zero. According to Maxwellian electrodynamics, the electromagnetic fields at a position $\boldsymbol{r}$ with respect to an origin inside the molecule where $r = |\boldsymbol{r}\,|$ is large compared to both the wavelength λ of the electromagnetic wave ($\lambda = 2\pi c/\omega$) and the diameter R of the molecule, are

$$\boldsymbol{E} = \boldsymbol{B} \times \hat{\boldsymbol{n}} \,, \tag{1.2}$$

and

$$\boldsymbol{B} = \frac{1}{c^2 r}[\ddot{\boldsymbol{d}}] \times \hat{\boldsymbol{n}} \,. \tag{1.3}$$

In Eqs. (1.2) and (1.3), $\hat{\boldsymbol{n}} = \boldsymbol{r}/r$ is a unit vector directed from the origin to the point of observation P and

$$[\boldsymbol{d}] = \boldsymbol{d}\left(t - \frac{r}{c}\right) \tag{1.4}$$

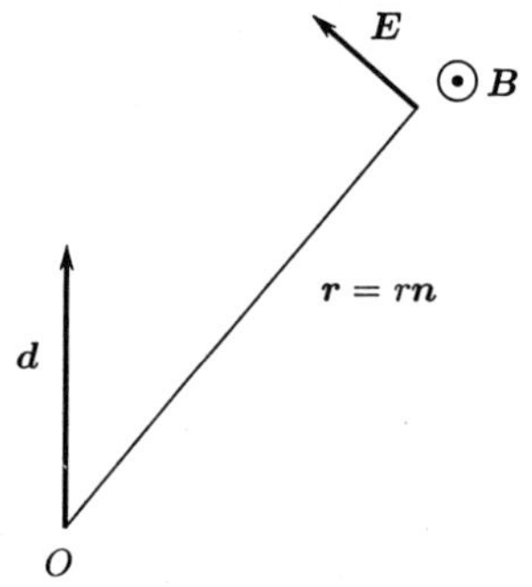

Figure 1.1 The electric and magnetic fields in the radiation region for an oscillating dipole. $\boldsymbol{B}$ is perpendicular to the plane of the figure and points toward the observer.

is the retarded value of the dipole moment at time t as seen from P. The directions of the $\boldsymbol{E}$ and $\boldsymbol{B}$ fields are shown schematically in Fig. 1. The magnitudes of the fields at P are

$$E = B = \frac{\omega^2 |\boldsymbol{d}_0|}{c^2 r} \sin\theta \cos\omega \left(t - \frac{r}{c}\right) . \tag{1.5}$$

We note that in the vicinity of the point P the electric and magnetic fields produced by (1.1) behave as a plane wave propagating along $\hat{\boldsymbol{n}}$ but with amplitudes proportional to $\sin\theta$, θ being the angle between $\boldsymbol{d}_0$ and $\hat{\boldsymbol{n}}$. Thus the intensity vanishes if $\overrightarrow{OP}$ is along $\boldsymbol{d}_0$ and has a maximum in the equatorial plane of the oscillating dipole. It is interesting to note that the field is always polarized in the plane through the point of observation containing the dipole moment.

The Poynting vector is

$$\boldsymbol{S} = \frac{c}{4\pi}\boldsymbol{E}\times\boldsymbol{B} = \frac{cB^2}{4\pi}\hat{\boldsymbol{n}} = \frac{\omega^4 |\boldsymbol{d}_0|^2 \cos^2\omega\left(t - \frac{r}{c}\right)\sin^2\theta}{4\pi c^3 r^2}\hat{\boldsymbol{n}} \tag{1.6}$$

The total power radiated is

$$P = \int \boldsymbol{S}\cdot\hat{\boldsymbol{n}} r^2 d\Omega = \frac{2}{3c^3}\omega^4 |\boldsymbol{d}_0|^2 \cos^2\omega\left(t - \frac{r}{c}\right) . \tag{1.7}$$

Averaging over a period of the oscillation we obtain the average power radiated, namely

$$\overline{P} = \frac{\omega^4 |\boldsymbol{d}_0|^2}{3c^3} . \tag{1.8}$$

This result was first obtained by Lord Rayleigh and used by him to explain the blue color of the sky. We remark that, according to this result, if the incident light is "white" rather than monochromatic, the higher rather than the lower frequencies are scattered more effectively. Thus the power spectrum of the scattered radiation is shifted with respect to that of the incident radiation toward the blue. This accounts for the blue color of the sky in directions away from the sun. In the same way this phenomenon explains the reddish color of the sunset. At the time of sunset the light from the sun reaching the observer traverses a larger thickness of air than at other hours. Light from the sun is scattered preferentially in the blue region of the spectrum and, therefore, the spectral density of the transmitted light is enriched toward the red.

These results also explain the partial polarization of sky light. Suppose we look at the sky in a direction at right angles to that of the incident sunlight. The sunlight is unpolarized and can, therefore, be regarded as composed of two plane polarized vibrations $\boldsymbol{E}_1$ and $\boldsymbol{E}_2$ at right angles to the direction of propagation and to each other and having no phase correlation. If $\boldsymbol{k}_0 = (2\pi/\lambda)\hat{\boldsymbol{n}}_0$ and $\boldsymbol{k} = (2\pi/\lambda)\hat{\boldsymbol{n}}$ are the propagation vectors of the incident and scattered beams, $\boldsymbol{E}_1$ may be taken perpendicular to the plane of $\boldsymbol{k}_0$ and $\boldsymbol{k}$ and $\boldsymbol{E}_2$ contained in it. The fields $\boldsymbol{E}_1$ and $\boldsymbol{E}_2$ induce oscillating electric dipole moments $\boldsymbol{d}_1 = \alpha\boldsymbol{E}_1$ and $\boldsymbol{d}_2 = \alpha\boldsymbol{E}_2$ on the air molecules which we suppose have an isotropic polarizability α. We note that the radiation originating from $\boldsymbol{d}_2$ has zero intensity in the direction of observation $\hat{\boldsymbol{n}}$ and hence, the scattered radiation in this direction is polarized in a plane normal to $\hat{\boldsymbol{n}}_0$. We will not discuss the partial polarization in other directions and that due to the anisotropy of the polarizability of the molecules.[1]

Probably the earliest scientific description concerning the scattering of light was given by Leonardo da Vinci.[2] He recorded the familiar fact that distant mountains have a blue tint that they do not exhibit when looked at in their immediate vicinity. He correctly attributed this effect to the presence of air between the mountain and the observer. The sunlight reflected by the mountain passes by the air layers between mountain and observer. The light from the sun is scattered by the air producing a lateral beam whose frequency spectrum is shifted toward the blue. The intensity of this beam is sufficient to compensate the shift toward the red of the reflected ray from the mountain.

The light scattered by the individual molecules of a gas or liquid is composed of beams having no phase correlations. Thus, if the molecules were uniformly distributed, the total intensity of the scattered radiation would vanish. Because of their motion, fluctuations in molecular density arise which, in turn produce a non–vanishing average intensity of scattered radiation. Let us suppose the scattering system is composed of N identical molecules. Denoting by $\boldsymbol{E}_i$ the electric field intensity at a point P due to scattering of the *ith* molecule, the total field at P is

$$\boldsymbol{E} = \sum_{i=1}^{N} \boldsymbol{E}_i \, . \tag{1.9}$$

We suppose P to be far from the scattering system. The mean square value $\overline{E^2}$ at P is

$$\overline{E^2} = \overline{\left(\sum_{i=1}^{N} \boldsymbol{E}_i\right)^2} = N\overline{E_i^2} \, , \tag{1.10}$$

the average of the cross terms vanishing since the phases are uncorrelated. The total power scattered per unit volume of fluid composed of molecules of polarizability α is

$$P = \frac{\omega^4}{3c^3}\,\alpha^2 n E_0^2 \tag{1.11}$$

where E_0 is the amplitude of the incident beam and n the density of molecules.

This result allows us to calculate the attenuation of light due to scattering. Let $I(x)$ be the energy flux of a beam of light of angular frequency ω propagating along the x–axis at the cross section characterized by the coordinate x. For a small increment Δx of x

$$I(x + \Delta x) = I(x) - P\Delta x \tag{1.12}$$

so that

$$\frac{dI}{dx} = -P = -\gamma\frac{cE^2}{8\pi} = -\gamma I(x) \tag{1.13}$$

with

$$\gamma = \frac{8\pi}{3}\left(\frac{\omega}{c}\right)^4 n\alpha^2 \ . \tag{1.14}$$

Thus,

$$I(x) = I(0)e^{-\gamma x} \ , \tag{1.15}$$

where γ is the attenuation coefficient. A measurement of γ, giving $n\alpha^2$, coupled with a measurement of the dielectric constant ϵ allowed an early determination of molecular polarizabilities and of Avogadro's number. To show that this is possible we remember that ϵ is related to $n\alpha$ by the Clausius–Mossotti relation

$$\frac{\epsilon - 1}{\epsilon + 2} = \frac{4\pi}{3} n\alpha \ . \tag{1.16}$$

It is not surprising, in view of the above, that scattering by gases is stronger than that by liquids and solids having the same number of molecules. The large fluctuations occurring in the vicinity of a phase transition give rise to scattering so strong that some otherwise transparent liquids become opaque under such circumstances. This phenomenon, which receives the name of critical opalescence, will be studied later.

So far we have considered that the scattered radiation occurs at the same frequency as the incident light. If one analyzes the scattered radiation emitted by matter in the presence of a monochromatic beam of angular frequency ω_0, he finds, in general that the scattered radiation consists of a beam of frequency ω_0 accompanied by weaker lines of frequencies $\omega_0 \pm \omega_i$. It is found that the frequency shifts ω_i are characteristic of the scattering medium and independent of ω_0. This effect is, therefore, different from the phenomenon of luminescence in which the frequency of the emitted radiation is independent of ω_0. The effect just described is called the Raman effect after its discoverer.[3] It had been predicted by Smekal[4] and its theory was implicit in the early work of Kramers and Heisenberg.[5] The effect was observed by Landsberg and Mandel'shtam[6] later but independently of

Raman. The phenomenon just described can be understood quantum mechanically as inelastic scattering of light accompanied by the excitation or by the absorption of a quantum of an internal mode of motion of the scattering system. In the process of inelastic scattering of light the scattering system is excited to a level of energy, say, $\hbar\omega_i$ above the ground state so that if ω_0 is the angular frequency of the incident light then that of the scattered radiation is $\omega' = \omega_0 - \omega_i$. The line at $\omega' = \omega_0 - \omega_i$ is called a Stokes line because Stokes, in his studies of luminescence, pointed out that the wavelength of the luminescent radiation is always longer than that of the incident radiation. If the scattering system is initially in an excited state and loses the energy $\hbar\omega_i$ then the scattered photon has energy $\hbar\omega' = \hbar(\omega_0 + \omega_i)$. A line at position $\omega_0 + \omega_i$ in a Raman spectrum is called an anti–Stokes line. Rayleigh scattering is that for which $\omega' = \omega_0$, *i.e.*, elastic scattering.

To understand this phenomenon from the classical point of view we imagine a molecule having internal motions characterized by normal coordinates

$$Q_i = Q_{i0} \cos \omega_i t \; . \tag{1.17}$$

To fix the ideas we view the coordinates Q_i as representing internal vibrational motions of the molecule. Only if $\omega_i < \omega_0$ is it possible to excite such modes. The polarizability α of the molecule is a function of the Q_i slowly modulating its value. Thus, we write

$$\boldsymbol{d} = \alpha(Q_1, Q_2, \ldots)\boldsymbol{E} \; . \tag{1.18}$$

Letting α_0 be the polarizability in the absence of molecular vibrations and defining

$$\alpha_i' = \left(\frac{\partial \alpha}{\partial Q_i}\right)_{Q_1=\ldots=Q_N=0} , \qquad \alpha_{ij}'' = \left(\frac{\partial^2 \alpha}{\partial Q_i \partial Q_j}\right)_{Q_1=\ldots=Q_N=0}$$

we obtain

$$\alpha = \alpha_0 + \sum_i \alpha_i' Q_i + \tfrac{1}{2} \sum_{ij} \alpha_{ij}'' Q_i Q_j + \ldots \tag{1.19}$$

for sufficiently small amplitudes Q_{i0}. We immediately find that the induced electric dipole moment has the form

$$\begin{aligned} d = {} & \alpha_0 E_0 \cos \omega_0 t + \frac{1}{2} E_0 \sum_i \alpha_i' Q_{i0} [\cos(\omega_0 + \omega_i)t + \cos(\omega_0 - \omega_i)t] \\ & + \frac{1}{8} E_0 \sum_{ij} \alpha_{ij}'' Q_{i0} Q_{j0} [\cos(\omega_0 + \omega_i + \omega_j)t + \cos(\omega_0 + \omega_i - \omega_j)t \\ & \qquad + \cos(\omega_0 - \omega_i + \omega_j)t + \cos(\omega_0 - \omega_i - \omega_j)\, t] \\ & + \cdots \end{aligned} \tag{1.20}$$

The first term in Eq. (1.20) accounts for Rayleigh scattering; the second corresponds to the anti–Stokes and Stokes Raman lines; the third is associated with the so–called second order Raman effect in which the scattering is accompanied by the emission and/or absorption of two quanta of molecular vibration. Higher order terms, though present, are rarely of interest. This theory is inadequate since it predicts equal intensities for Stokes and anti–Stokes lines in the presence of internal motions and does not account for the Raman effect when the molecule is in its lowest energy state. This, as well as the consequences of the symmetry of the scattering system, is best discussed within the framework of the quantum theory. Of course, a semiclassical theory, similar to that developed by Einstein, using the correspondence principle, for absorption and emission of radiation can also be given. Before presenting the quantum theory of the scattering of light we review the quantization of the electromagnetic field.

Chapter 2

Expansion of the Electromagnetic Field in a Cavity in Plane Waves

The normal modes of mechanical and electromagnetic systems are defined so that the ratios of amplitudes at different positions are independent of time, i.e., they are standing waves. Use of the normal modes in subsequent development leads to mathematical complications when analyzing the interaction of electromagnetic waves with material systems. This is because the law of conservation of momentum is more conveniently expressed in terms of traveling waves rather than in terms of standing waves which do not possess well defined momenta.

Therefore, we consider a cubic cavity of side L but instead of assuming that its walls are perfectly reflecting we suppose that the fields are periodic with the cube as the fundamental period. It is enough to take the vector potential $\boldsymbol{A}(\boldsymbol{r},t)$ as a periodic function, i.e., we suppose that

$$\boldsymbol{A}(\boldsymbol{r}+n_xL\hat{\boldsymbol{x}}+n_yL\hat{\boldsymbol{y}}+n_zL\hat{\boldsymbol{z}},t)=\boldsymbol{A}(\boldsymbol{r},t) \tag{2.1}$$

for $n_x, n_y, n_z = 0, \pm1, \pm2, \ldots$. We argue that, from the physical standpoint, this assumption is equivalent to that of perfectly reflecting walls. In fact, the purpose of the walls is merely to keep the electromagnetic energy confined within the cavity. The same result is accomplished by the use of periodic boundary conditions because any amount of electromagnetic energy leaving through one face of the cube is compensated exactly by an equal amount entering through the opposite face.

With this in mind, any triply periodic field with periods L in x, y and z can be expanded in a Fourier series. The expansion of the vector potential is

$$\boldsymbol{A}(\boldsymbol{r},t)=\sum_{k}\boldsymbol{A}_k(t)\exp(i\boldsymbol{k}\cdot\boldsymbol{r})\ . \tag{2.2}$$

Here the sum over $\boldsymbol{k}$ extends over all vectors whose components are $k_i=(2\pi/L)n_i$; $i = x, y, z$; $n_i = 0, \pm 1, \pm 2, \ldots$. The functions $\exp i\boldsymbol{k}\cdot\boldsymbol{r}$ are orthogonal, i.e.,

$$\int_{(V)} \exp(-i\boldsymbol{k}'\cdot\boldsymbol{r})\exp(i\boldsymbol{k}\cdot\boldsymbol{r})d\boldsymbol{r} = \delta_{k'k}V \tag{2.3}$$

where $V = L^3$. The coefficients $\boldsymbol{A}_k(t)$ are called the Fourier coefficients of the vector potential. Since $\boldsymbol{A}(\boldsymbol{r},t)$ is real,

$$\boldsymbol{A}_{-k}(t) = \boldsymbol{A}_k^*(t) \ . \tag{2.4}$$

In addition, in the absence of electric charges we can select $\boldsymbol{A}(\boldsymbol{r},t)$ so that $\nabla\cdot\boldsymbol{A} = 0$. Then

$$\boldsymbol{k}\cdot\boldsymbol{A}_k(t) = 0 \ , \tag{2.5}$$

i.e., $\boldsymbol{A}_k(t)$ is orthogonal to $\boldsymbol{k}$ and, hence, it has two independent components. These correspond to the two possible independent polarizations of plane waves propagating with wave vector $\boldsymbol{k}$. For each $\boldsymbol{k}$ we take two unit vectors $\hat{\boldsymbol{e}}_{k1}$ and $\hat{\boldsymbol{e}}_{k2}$ orthogonal to each other and such that

$$\hat{\boldsymbol{e}}_{k1}\times\hat{\boldsymbol{e}}_{k2} = \hat{\boldsymbol{k}} = \boldsymbol{k}/|\boldsymbol{k}| \ . \tag{2.6}$$

The wave equation shows that the Fourier coefficient $\boldsymbol{A}_k(t)$ satisfies the differential equation

$$\ddot{\boldsymbol{A}}_k(t) = -c^2k^2\boldsymbol{A}_k(t) = -\omega_k^2\boldsymbol{A}_k(t) \tag{2.7}$$

where

$$\omega_k = c|\boldsymbol{k}| \ . \tag{2.8}$$

Thus, each Fourier component performs a harmonic vibration with angular frequency $\omega_k = c|\boldsymbol{k}|$.

The electric and magnetic fields expressed in the form of Fourier series are

$$\boldsymbol{E}(\boldsymbol{r},t) = -\frac{1}{c}\partial_t\boldsymbol{A} = -\frac{1}{c}\sum_k \dot{\boldsymbol{A}}_k(t)\exp(i\boldsymbol{k}\cdot\boldsymbol{r}) \tag{2.9}$$

and

$$\boldsymbol{B}(\boldsymbol{r},t) = \nabla\times\boldsymbol{A} = i\sum_k \boldsymbol{k}\times\boldsymbol{A}_k(t)\exp(i\boldsymbol{k}\cdot\boldsymbol{r}) \ . \tag{2.10}$$

The electromagnetic energy within the volume of the cube is

$$U = \frac{1}{8\pi}\int_{(V)}(E^2+B^2)d\boldsymbol{r} = \frac{V}{8\pi c^2}\sum_k\left(\dot{\boldsymbol{A}}_k\cdot\dot{\boldsymbol{A}}_{-k} + \omega_k^2\boldsymbol{A}_k\cdot\boldsymbol{A}_{-k}\right) \tag{2.11}$$

where we made use of Eq. (2.5) and of the orthogonality of the functions $\{\exp(i\boldsymbol{k}\cdot\boldsymbol{r})\}$ expressed in Eq. (2.3).

The general solution of Eq. (2.7) is

$$\boldsymbol{A}_k(t) = \boldsymbol{A}_k(0)\cos\omega_k t + \omega_k^{-1}\dot{\boldsymbol{A}}_k(0)\sin\omega_k t$$

where $\boldsymbol{A}_k(0)$ and $\dot{\boldsymbol{A}}_k(0)$ are the values of $\boldsymbol{A}_k(t)$ and its time derivative at $t = 0$, i.e., the initial conditions. We can rewrite this solution using the complex quantity

$$\boldsymbol{a}_k(t) = \tfrac{1}{2}\left[\boldsymbol{A}_k(0) + i\omega_k^{-1}\dot{\boldsymbol{A}}_k(0)\right]\exp(-i\omega_k t)$$

and its complex conjugate. Then

$$\boldsymbol{A}_k(t) = \boldsymbol{a}_k(t) + \boldsymbol{a}^*_{-k}(t) \; . \tag{2.12}$$

The advantage of the variables $\boldsymbol{a}_k(t)$ over the Fourier components $\boldsymbol{A}_k(t)$ is that they satisfy the simple first order differential equation

$$\dot{\boldsymbol{a}}_k(t) = -i\omega_k \boldsymbol{a}_k(t) \; . \tag{2.13}$$

Furthermore, the condition (2.4) requiring $\boldsymbol{A}(\boldsymbol{r}, t)$ to be real is automatically satisfied by Eq. (2.12).

The energy U and the Poynting vector are

$$U = \frac{V}{2\pi c^2}\sum_k \omega_k^2 \boldsymbol{a}_k^* \cdot \boldsymbol{a}_k \tag{2.14}$$

and

$$\boldsymbol{S} = \frac{c}{4\pi}\boldsymbol{E}\times\boldsymbol{B} = -\frac{i}{4\pi}\sum_{k,k'}\dot{\boldsymbol{A}}_k \times \left(\boldsymbol{k}' \times \boldsymbol{A}_{k'}\right)\exp[i(\boldsymbol{k}'+\boldsymbol{k})\cdot\boldsymbol{r}] \tag{2.15}$$

so that the electromagnetic momentum within the cube is

$$\boldsymbol{\Pi} = \frac{1}{c^2}\int_{(V)} \boldsymbol{S} d\boldsymbol{r} = \frac{V}{2\pi c^2}\sum_k \boldsymbol{k}\omega_k \boldsymbol{a}_k^* \cdot \boldsymbol{a}_k \; . \tag{2.16}$$

These results can be interpreted as follows. Each wave vector $\boldsymbol{k}$ contributes

$$u_k = \frac{V}{2\pi c^2}\omega_k^2 \boldsymbol{a}_k^* \cdot \boldsymbol{a}_k \tag{2.17}$$

to the energy and

$$\boldsymbol{p}_k = \frac{V}{2\pi c^2}\boldsymbol{k}\omega_k \boldsymbol{a}_k^* \cdot \boldsymbol{a}_k = \frac{u_k}{c}\frac{\boldsymbol{k}}{|\boldsymbol{k}|} \tag{2.18}$$

to the momentum in the cavity.

We now introduce the real variables

$$\boldsymbol{Q}_k = \frac{1}{c}\left(\frac{V}{4\pi}\right)^{1/2}(\boldsymbol{a}_k + \boldsymbol{a}_k^*) \tag{2.19}$$

having two linearly independent components perpendicular to the wave vector $\boldsymbol{k}$. The time derivative of $\boldsymbol{Q}_k$ is denoted by $\boldsymbol{P}_k$ and is equal to

$$\boldsymbol{P}_k = \dot{\boldsymbol{Q}}_k = -\frac{i\omega_k}{c}\left(\frac{V}{4\pi}\right)^{1/2}(\boldsymbol{a}_k - \boldsymbol{a}_k^*) \; . \tag{2.20}$$

Solving Eqs. (2.19) and (2.20) for $\boldsymbol{a}_k$ and substituting into Eqs. (2.14) and (2.16) we find

$$U = \frac{1}{2}\sum_k (P_k^2 + \omega_k^2 Q_k^2) \tag{2.21}$$

and

$$\boldsymbol{\Pi} = \frac{1}{2c}\sum_k (P_k^2 + \omega_k^2 Q_k^2)\frac{\boldsymbol{k}}{|\boldsymbol{k}|} \ . \tag{2.22}$$

These results show that we can describe the radiation field as a collection of two dimensional harmonic oscillators, one for each value of the wave vector $\boldsymbol{k}$. The two components in the displacement of each oscillator correspond to the linearly independent polarizations of a plane wave with propagation vector $\boldsymbol{k}$. The analogy between the electromagnetic field in the cavity and the mechanical system described by coordinates $\boldsymbol{Q}_k$ and momenta $\boldsymbol{P}_k$ is now complete. If in fact, we regard U as the Hamiltonian function for the system, the equations of motion are

$$\dot{\boldsymbol{Q}}_k = \frac{\partial U}{\partial \boldsymbol{P}_k} = \boldsymbol{P}_k \ , \quad \dot{\boldsymbol{P}}_k = -\frac{\partial U}{\partial \boldsymbol{Q}_k} = -\omega_k^2 \boldsymbol{Q}_k$$

so that

$$\ddot{\boldsymbol{Q}}_k = -\omega_k^2 \boldsymbol{Q}_k \ .$$

But this is exactly the equation for $\boldsymbol{A}_k(t)$ obtained by differentiation of Eq. (2.12).

The vector potential in terms of $\boldsymbol{Q}_k$ and $\boldsymbol{P}_k$ is

$$\begin{aligned}\boldsymbol{A}(\boldsymbol{r},t) &= \sum_k [\boldsymbol{a}_k(t)\exp(i\boldsymbol{k}\cdot\boldsymbol{r}) + \boldsymbol{a}_k^*(t)\exp(-i\boldsymbol{k}\cdot\boldsymbol{r})] \\ &= \left(\frac{4\pi c^2}{V}\right)^{1/2}\sum_k \left[\boldsymbol{Q}_k(t)\cos\boldsymbol{k}\cdot\boldsymbol{r} - \frac{1}{\omega_k}\boldsymbol{P}_k(t)\sin\boldsymbol{k}\cdot\boldsymbol{r}\right] \ .\end{aligned} \tag{2.23}$$

The electric and magnetic fields are

$$\boldsymbol{E}(\boldsymbol{r},t) = -\frac{1}{c}\partial_t \boldsymbol{A}(\boldsymbol{r},t) = -\left(\frac{4\pi}{V}\right)^{1/2}\sum_k [\omega_k \boldsymbol{Q}_k(t)\sin\boldsymbol{k}\cdot\boldsymbol{r} + \boldsymbol{P}_k(t)\cos\boldsymbol{k}\cdot\boldsymbol{r}] \ , \tag{2.24}$$

and

$$\boldsymbol{B}(\boldsymbol{r},t) = \nabla\times\boldsymbol{A} = -\left(\frac{4\pi}{V}\right)^{1/2}\sum_k \frac{\boldsymbol{k}}{|\boldsymbol{k}|}\times[\omega_k \boldsymbol{Q}_k(t)\sin\boldsymbol{k}\cdot\boldsymbol{r} + \boldsymbol{P}_k(t)\cos\boldsymbol{k}\cdot\boldsymbol{r}] \ . \tag{2.25}$$

The angular momentum in the field within the cavity is

$$\frac{1}{4\pi c}\int_{(V)} \boldsymbol{r}\times(\boldsymbol{E}\times\boldsymbol{B})d\boldsymbol{r} \ .$$

The i–component of this vector is

$$\frac{1}{4\pi c}\epsilon_{ijk}\int_{(V)} x_j(\boldsymbol{E}\times\boldsymbol{B})_k d\boldsymbol{r} = \frac{1}{4\pi c}\epsilon_{ijk}\int_{(V)} x_j(E_\ell\partial_k A_\ell - E_\ell\partial_\ell A_k)d\boldsymbol{r}$$
$$= \frac{1}{4\pi c}\epsilon_{ijk}\int_{(V)} x_j E_\ell\partial_k A_\ell d\boldsymbol{r} + \frac{1}{4\pi c}\epsilon_{ijk}\int_{(V)} E_j A_k d\boldsymbol{r}\ . \tag{2.26}$$

The second term has been transformed making use of $\nabla\cdot\boldsymbol{E}=0$ and an integration by parts; since it does not depend on the choice of origin we interpret it as the intrinsic angular momentum of the radiation. Thus we write

$$\boldsymbol{L} = \frac{1}{4\pi c}\int_{(V)} \boldsymbol{E}\times\boldsymbol{A}d\boldsymbol{r}\ . \tag{2.27}$$

Using Eqs. (2.2) and (2.9) we obtain

$$\boldsymbol{L} = \frac{V}{4\pi c^2}\sum_k \boldsymbol{A}_k^*\times\dot{\boldsymbol{A}}_k = -\frac{iV}{2\pi c^2}\sum_k \omega_k\boldsymbol{a}_k^*\times\boldsymbol{a}_k \tag{2.28}$$

where we used, in addition, Eqs. (2.12), (2.13) and

$$\sum_k \omega_k\boldsymbol{a}_{-k}\times\boldsymbol{a}_k = \sum_k \omega_k\boldsymbol{a}_k\times\boldsymbol{a}_{-k} = 0$$

and a similar expression involving the $\boldsymbol{a}_k^*$. Equation (2.28) can be transformed further with the aid of Eqs. (2.19) and (2.20) which yield

$$\boldsymbol{a}_k^*\times\boldsymbol{a}_k = \frac{2\pi i c^2}{V\omega_k}\boldsymbol{Q}_k\times\boldsymbol{P}_k\ .$$

Thus

$$\boldsymbol{L} = \sum_k \boldsymbol{Q}_k\times\boldsymbol{P}_k\ . \tag{2.29}$$

We conclude that the radiation in the cavity behaves as a collection of two–dimensional simple harmonic oscillators, one for each value of

$$\boldsymbol{k} = \frac{2\pi}{L}(n_1,n_2,n_3)\ ,\qquad n_i = 0,\pm1,\pm2,\dots\ .$$

The frequency is

$$\omega_k = c|\boldsymbol{k}| = \left[\frac{4\pi^2}{L^2}(n_1^2+n_2^2+n_3^2)c^2\right]^{1/2}\ .$$

The number of **one-dimensional oscillators** having frequency less than ω is

$$2(4\pi/3)(\omega/c)^3\frac{L^3}{(2\pi)^3} = \frac{V\omega^3}{3\pi^2c^3}\ .$$

Thus, the density of states, i.e., the number of one–dimensional oscillators per unit volume and per unit frequency range at ω is

$$g(\omega) = \frac{\omega^2}{\pi^2c^3}\ . \tag{2.30}$$

Chapter 3

Quantization of the Electromagnetic Field

In Chapter 2, a relation was established between the Fourier components of the vector potential of the free radiation field and a set of one–dimensional harmonic oscillators. The energy in the field is described by the Hamiltonian function

$$H = \tfrac{1}{2} \sum_{k\mu} (P_{k\mu}^2 + \omega_k^2 Q_{k\mu}^2) \tag{3.1}$$

where $\mu = 1, 2$ distinguishes between the two possible polarizations of a wave with wave vector $\boldsymbol{k}$. The coordinates introduced in Chapter 2 are

$$\boldsymbol{Q}_k = Q_{k1}\hat{\mathbf{e}}_{k1} + Q_{k2}\hat{\mathbf{e}}_{k2} \tag{3.2}$$

where $\hat{\mathbf{e}}_{k1}$ and $\hat{\mathbf{e}}_{k2}$ are unit vectors orthogonal to $\hat{\boldsymbol{k}} = \boldsymbol{k}/|\boldsymbol{k}|$ and to each other ($\hat{\mathbf{e}}_{k1} \times \hat{\mathbf{e}}_{k2} = \hat{\boldsymbol{k}}$). The two-dimensional momentum $\boldsymbol{P}_k$ is given by an expression similar to (3.2).

We saw that the classical electric and magnetic fields are derived in the radiation gauge from the solenoidal vector potential

$$\boldsymbol{A}(\boldsymbol{r}, t) = \left(\frac{4\pi c^2}{V}\right)^{1/2} \sum_{k\mu} \hat{\mathbf{e}}_{k\mu}(Q_{k\mu} \cos \boldsymbol{k}\cdot\boldsymbol{r} - (P_{k\mu}/\omega_k) \sin \boldsymbol{k}\cdot\boldsymbol{r}) \; . \tag{3.3}$$

The electric and magnetic fields are

$$\boldsymbol{E}(\boldsymbol{r}, t) = -\left(\frac{4\pi}{V}\right)^{1/2} \sum_{k\mu} \hat{\mathbf{e}}_{k\mu}(\omega_k Q_{k\mu} \sin \boldsymbol{k}\cdot\boldsymbol{r} + P_{k\mu} \cos \boldsymbol{k}\cdot\boldsymbol{r}) \tag{3.4}$$

and

$$\boldsymbol{B}(\boldsymbol{r}, t) = -\left(\frac{4\pi}{V}\right)^{1/2} \sum_{k\mu} \hat{\boldsymbol{k}} \times \hat{\mathbf{e}}_{k\mu}(\omega_k Q_{k\mu} \sin \boldsymbol{k}\cdot\boldsymbol{r} + P_{k\mu} \cos \boldsymbol{k}\cdot\boldsymbol{r}) \; , \tag{3.5}$$

respectively.

To quantize the free electromagnetic field we proceed according to the following rules. We replace the classical variables $Q_{k\mu}$ and $P_{k\mu}$ by operators obeying the commutation relations

$$[Q_{k'\mu'}, Q_{k\mu}] = 0 \ , \quad [P_{k'\mu'}, P_{k\mu}] = 0 \ , \quad [Q_{k'\mu'}, P_{k\mu}] = i\hbar\delta_{k'k}\delta_{\mu'\mu} \ . \tag{3.6}$$

We conclude then that the Hamiltonian operator

$$H_{k\mu} = \frac{1}{2}(P_{k\mu}^2 + \omega_k^2 Q_{k\mu}^2) \tag{3.7}$$

has the discrete spectrum characterized by the eigenvalues

$$E_{k\mu} = \hbar\omega_k\left(n_{k\mu} + \frac{1}{2}\right) , \tag{3.8}$$

$n_{k\mu} = 0, 1, 2, \ldots$. The eigenvectors, denoted by $|\{n_{k\mu}\}\rangle$, and eigenvalues of H are described by a set of occupation numbers $n_{k\mu}$ for all choices of $\boldsymbol{k}$ μ. The corresponding eigenvalues for the radiation field are

$$\sum_{k\mu} \hbar\omega_k\left(n_{k\mu} + \frac{1}{2}\right) . \tag{3.9}$$

The infinite zero point energy $\Sigma(\hbar\omega_k/2)$ is a constant and need not concern us here.

We now introduce the raising and lowering operators $a_{k\mu}^\dagger$ and $a_{k\mu}$ as is done in the discussion of the simple harmonic oscillator. We write

$$a_{k\mu} = \left(\frac{\omega_k}{2\hbar}\right)^{1/2}\left(Q_{k\mu} + \frac{iP_{k\mu}}{\omega_k}\right) \tag{3.10}$$

The operator $a_{k\mu}^\dagger$ is

$$a_{k\mu}^\dagger = \left(\frac{\omega_k}{2\hbar}\right)^{1/2}\left(Q_{k\mu} - \frac{iP_{k\mu}}{\omega_k}\right) . \tag{3.11}$$

From Eqs. (3.10), (3.11) and (3.6) we conclude that

$$[a_{k'\mu'}, a_{k\mu}] = 0 \ , \quad \left[a_{k'\mu'}, a_{k\mu}^\dagger\right] = \delta_{k'k}\delta_{\mu'\mu} \ . \tag{3.12}$$

The classical vector fields $\boldsymbol{A}$, $\boldsymbol{E}$ and $\boldsymbol{B}$ are functions of position and time by virtue of the time dependence of the coordinates and momenta, but the operators $\boldsymbol{A}$, $\boldsymbol{E}$ and $\boldsymbol{B}$ are time-independent in the Schrödinger representation. They can be expressed in terms of the raising and lowering operators by solving Eqs. (3.10) and (3.11) for $Q_{k\mu}$ and $P_{k\mu}$ in terms of $a_{k\mu}$ and $a^\dagger_{k\mu}$ and substituting in Eqs. (3.3)–(3.5). We obtain

$$\boldsymbol{A}(\boldsymbol{r}) = \sum_{k\mu} \left(\frac{2\pi\hbar c}{Vk}\right)^{1/2} \hat{\boldsymbol{e}}_{k\mu} \left[a_{k\mu} \exp(i\boldsymbol{k}\cdot\boldsymbol{r}) + a^\dagger_{k\mu} \exp(-i\boldsymbol{k}\cdot\boldsymbol{r})\right] , \tag{3.13}$$

$$\boldsymbol{E}(\boldsymbol{r}) = i\sum_{k\mu} \left(\frac{2\pi\hbar\omega_k}{V}\right)^{1/2} \hat{\boldsymbol{e}}_{k\mu} \left[a_{k\mu} \exp(i\boldsymbol{k}\cdot\boldsymbol{r}) - a^\dagger_{k\mu} \exp(-i\boldsymbol{k}\cdot\boldsymbol{r})\right] \tag{3.14}$$

and

$$\boldsymbol{B}(\boldsymbol{r}) = i\sum_{k\mu} \left(\frac{2\pi\hbar\omega_k}{V}\right)^{1/2} \hat{\boldsymbol{k}} \times \hat{\boldsymbol{e}}_{k\mu} \left[a_{k\mu} \exp(i\boldsymbol{k}\cdot\boldsymbol{r}) - a^\dagger_{k\mu} \exp(-i\boldsymbol{k}\cdot\boldsymbol{r})\right] . \tag{3.15}$$

The operators $a_{k\mu}$ and $a^\dagger_{k\mu}$ are called the destruction and creation operators for photons characterized by wave vector $\boldsymbol{k}$ and polarization μ, respectively. The operators $a_{k\mu}$ and $a^\dagger_{k\mu}$ upon acting on a stationary state $|\{n_{k\mu}\}\rangle$ of the system, reduce and increase the photon occupation number $n_{k\mu}$ in the mode $\boldsymbol{k}\mu$ by one unity, i.e.,

$$a_{k\mu}|\{n_{k'\mu'}\}\rangle = (n_{k\mu})^{1/2}|\{n_{k'\mu'} - \delta_{k'k}\delta_{\mu'\mu}\}\rangle \tag{3.16}$$

and

$$a^\dagger_{k\mu}|\{n_{k'\mu'}\}\rangle = (n_{k\mu}+1)^{1/2}|\{n_{k'\mu'} + \delta_{k'k}\delta_{\mu'\mu}\}\rangle . \tag{3.17}$$

We conclude that the expectation values of $\boldsymbol{A}$, $\boldsymbol{E}$ and $\boldsymbol{B}$ in any stationary state of H vanish. This is not surprising. We recall that $\langle x\rangle$ and $\langle p\rangle$ vanish for any stationary state of a simple harmonic oscillator.

Chapter 4

Emission and Absorption of Electromagnetic Radiation

The Hamiltonian of a system of N charged particles of masses and charges m_i and q_i in a field of radiation characterized by a vector potential $\boldsymbol{A}(\boldsymbol{r})$ is

$$H = \sum_{i=1}^{N} \frac{1}{2m_i}\left(\boldsymbol{p}_i - \frac{q_i}{c}\boldsymbol{A}(\boldsymbol{r}_i)\right)^2 + U - \sum_{i=1}^{N} \boldsymbol{\mu}_i \cdot (\nabla \times \boldsymbol{A}(\boldsymbol{r}_i)) + H_R \,, \tag{4.1}$$

where $\boldsymbol{\mu}_i$ is the intrinsic magnetic moment of the *ith* particle, which we express in the form

$$\boldsymbol{\mu}_i = g_i \frac{q_i \hbar}{2m_i c} \boldsymbol{S}_i \,. \tag{4.2}$$

$H_R = \sum_{k\mu} \hbar\omega_k a^{\dagger}_{k\mu} a_{k\mu}$ is the Hamiltonian of the radiation field. The remaining symbols have the usual meanings. We emphasize that $\boldsymbol{A}(\boldsymbol{r})$ is the operator vector potential discussed in Chapter 3. We write

$$H = H_o + H'_I + H'_{II} = H_o + H' \,, \tag{4.3}$$

where

$$H_o = \sum_i \frac{p_i^2}{2m_i} + U + H_R = H_M + H_R \,, \tag{4.4}$$

H_M being the Hamiltonian of the material system in the absence of the electromagnetic field,

$$H'_I = -\sum_{i=1}^{N} \frac{q_i}{2m_i c}(\boldsymbol{A}(\boldsymbol{r}_i) \cdot \boldsymbol{p}_i + \boldsymbol{p}_i \cdot \boldsymbol{A}(\boldsymbol{r}_i)) - \sum_{i=1}^{N} \boldsymbol{\mu}_i \cdot (\nabla \times \boldsymbol{A}(\boldsymbol{r}_i)) \tag{4.5}$$

and

$$H'_{II} = \sum_i \frac{q_i^2}{2m_i c^2}(\boldsymbol{A}(\boldsymbol{r}_i))^2 \,. \tag{4.6}$$

We express $\boldsymbol{A}(\boldsymbol{r})$ in terms of the creation and destruction operators $a_{k\mu}^{\dagger}$ and $a_{k\mu}$ introduced in Chapter 3.

$$H_I' = -\sum_{k\mu}\left(\frac{2\pi\hbar}{V\omega_k}\right)^{1/2}\hat{e}_{k\mu}\cdot\left[a_{k\mu}\boldsymbol{J}^{\dagger}(\boldsymbol{k})+a_{k\mu}^{\dagger}\boldsymbol{J}(\boldsymbol{k})\right], \tag{4.7}$$

with

$$\boldsymbol{J}(\boldsymbol{k})=\sum_i\left[\frac{q_i}{2m_i}(\boldsymbol{p}_i\exp(-i\boldsymbol{k}\cdot\boldsymbol{r}_i)+\exp(-i\boldsymbol{k}\cdot\boldsymbol{r}_i)\boldsymbol{p}_i)+ic\boldsymbol{k}\times\boldsymbol{\mu}_i\exp(-i\boldsymbol{k}\cdot\boldsymbol{r}_i)\right]. \tag{4.8}$$

The operator $\boldsymbol{J}(\boldsymbol{k})$ is the Fourier transform of the current density operator $\boldsymbol{J}(\boldsymbol{r})$ defined by

$$\boldsymbol{J}(\boldsymbol{r})=\sum_i\frac{q_i}{2m_i}\left(\boldsymbol{p}_i\delta(\boldsymbol{r}-\boldsymbol{r}_i)+\delta(\boldsymbol{r}-\boldsymbol{r}_i)\boldsymbol{p}_i\right)+c\nabla\times\boldsymbol{M} \tag{4.9}$$

where $\boldsymbol{M}(\boldsymbol{r})$ is the magnetization operator

$$\boldsymbol{M}(\boldsymbol{r})=\sum_i\boldsymbol{\mu}_i\delta(\boldsymbol{r}-\boldsymbol{r}_i). \tag{4.10}$$

The relations between $\boldsymbol{J}(\boldsymbol{r})$ and $\boldsymbol{J}(\boldsymbol{k})$ are

$$\boldsymbol{J}(\boldsymbol{r})=\frac{1}{V}\sum_k\boldsymbol{J}(\boldsymbol{k})\exp(i\boldsymbol{k}\cdot\boldsymbol{r}),\quad \boldsymbol{J}(\boldsymbol{k})=\int_{(V)}d\boldsymbol{r}\boldsymbol{J}(\boldsymbol{r})\exp(-i\boldsymbol{k}\cdot\boldsymbol{r}). \tag{4.11}$$

We note that, since $\hat{\boldsymbol{e}}_{k\mu}$ is perpendicular to $\boldsymbol{k}$, $\hat{\boldsymbol{e}}_{k\mu}\cdot\boldsymbol{p}_i$ and $\exp(-i\boldsymbol{k}\cdot\boldsymbol{r}_i)$ always commute. Clearly,

$$\hat{\boldsymbol{e}}_{k\mu}\cdot\boldsymbol{J}^{\dagger}(\boldsymbol{k})=\hat{\boldsymbol{e}}_{k\mu}\cdot\boldsymbol{J}(-\boldsymbol{k}). \tag{4.12}$$

To calculate the probability of absorption or emission of a photon we determine the evolution of the system consisting of the radiation field, the material system and their mutual interaction using the Schrödinger equation $i\hbar(\partial\psi/\partial t)=(H_0+H')\psi$. We suppose that at the initial time $t=0$ the system is known to be in one of the eigenstates of H_o. We designate the states of H_M by the indices ν and their energies by E_ν. The energy eigenvalues of H_R are $\sum_{k\mu}\hbar\omega_k(n_{k\mu}+\frac{1}{2})$. They will be denoted by r, a symbol representing the collection of occupation numbers $\{n_{k\mu}\}$. Let the initial state of the system be $|r_o\nu_o\rangle$. The probability that at time t the system is in $|r\nu\rangle$ is

$$\begin{aligned}P_{r_0\nu_0\to r\nu}(t)&=|\langle r\nu|\psi_t\rangle|^2\\&=\left|\left\langle r\nu\left|1+\sum_{n=1}^{\infty}(-i/\hbar)^n\int_0^t dt_n\int_0^{t_n}dt_{n-1}\ldots\int_0^{t_2}dt_1H'(t_n)H'(t_{n-1})\ldots H'(t_1)\right|r_0\nu_0\right\rangle\right|^2,\end{aligned} \tag{4.13}$$

where $H'(t) = \exp(iH_0t/\hbar)H'\exp(-iH_ot/\hbar)$ is the expression of the perturbation H' in the interaction representation. Let us suppose now that r_o and r differ by a single photon, say in the mode $k\mu$. Then, to first order in the interaction

$$P_{r_0\nu_0\to r\nu}(t) = \frac{1}{\hbar^2}\left|\int_0^t dt'\langle r\nu|H_I'(t')|r_0\nu_0\rangle\right|^2 . \tag{4.14}$$

If r_o has one more (less) photon than r, an absorption (emission) process has occurred. We have

$$\langle r\nu|H_I'(t')|r_0\nu_0\rangle = -\left(\frac{2\pi\hbar}{V\omega_k}\right)^{1/2}\hat{\boldsymbol{e}}_{k\mu}\cdot\langle\nu|\boldsymbol{J}^\dagger(\boldsymbol{k})|\nu_0\rangle(n_{k\mu})^{1/2}\exp[i(\omega_{\nu\nu_0}-\omega_k)]t', \tag{4.15}$$

for absorption and

$$\langle r\nu|H_I'(t')|r_0\nu_0\rangle = -\left(\frac{2\pi\hbar}{V\omega_k}\right)^{1/2}\hat{\boldsymbol{e}}_{k\mu}\cdot\langle\nu|\boldsymbol{J}(\boldsymbol{k})|\nu_0\rangle(n_{k\mu}+1)^{1/2}\exp[i(\omega_{\nu\nu_0}+\omega_k]t') , \tag{4.16}$$

for emission. Here $\hbar\omega_{\nu\nu_o} = E_\nu - E_{\nu_o}$. The probability that absorption of a photon $k\mu$ has occurred after time t is thus

$$P_{r_0\nu_0\to r\nu}(t) = \frac{2\pi n_{k\mu}}{V\hbar\omega_k}\left|\hat{\boldsymbol{e}}_{k\mu}\cdot\langle\nu|\boldsymbol{J}^\dagger(\boldsymbol{k})|\nu_0\rangle\right|^2 D(E_\nu - E_{\nu_0} - \hbar\omega_k) , \tag{4.17}$$

where $D(E)$ is $\sin^2(Et/2\hbar)/(E/2\hbar)^2$. The function $D(E)$ has a peak at $E = 0$ whose magnitude increases proportionally to t^2. The width of the peak is of order t^{-1} and subsidiary but smaller peaks occur at $E = (2n+1)\pi\hbar/t$, $n = 0, \pm1, \pm2, \ldots$. For sufficiently long times $D(E) \approx 2\pi\hbar t\delta(E)$. The probability per unit time that a transition from ν_o to ν with absorption of a photon whose wave vector is directed within the solid angle $d\Omega$ and having polarization $\hat{\boldsymbol{e}}_{k\mu}$ is

$$dw^{(\mu)}_{\nu_0\to\nu} = \frac{1}{t}\sum_{k\epsilon d\Omega}\frac{2\pi n_{k\mu}}{V\hbar\omega_k}\left|\hat{\boldsymbol{e}}_{k\mu}\cdot\left\langle\nu\middle|\boldsymbol{J}^\dagger(\boldsymbol{k})\middle|\nu_0\right\rangle\right|^2 D(E_\nu - E_{\nu_0} - \hbar\omega_k) . \tag{4.18}$$

Here we have assumed that the different photons have random phase relations with respect to one another so that we can add their probabilities rather than their probability amplitudes. The sum over $\boldsymbol{k}$ can be replaced by the integral

$$\sum_{k\epsilon d\Omega} \to \frac{V}{(2\pi)^3}\,d\Omega\int_0^\infty\frac{\omega^2 d\omega}{c^3} . \tag{4.19}$$

We define the spectral density $I_\mu(\omega)$ of the radiation with polarization μ by equating $I_\mu(\omega)d\omega d\Omega$ to the electromagnetic energy flux in the direction of $\boldsymbol{k}$ within the solid angle $d\Omega$ and having frequency in the range $d\omega$ at ω. We suppose $n_{k\mu} = n_\mu(\omega)$ is a slowly varying function of ω in this range. Then we have

$$I_\mu(\omega)d\omega d\Omega = \hbar\omega n_\mu(\omega)c\,\frac{1}{(2\pi)^3}\,\frac{\omega^2 d\omega d\Omega}{c^3} ,$$

$$I_\mu(\omega) = \frac{\hbar\omega^3 n_\mu(\omega)}{8\pi^3 c^2} \ . \tag{4.20}$$

Then

$$dw^{(\mu)}_{\nu_o\to\nu} = \frac{2\pi d\Omega}{\hbar^2 ct}\int_o^\infty \frac{I_\mu(\omega)d\omega}{\omega^2}\left|\hat{\boldsymbol{e}}_{k\mu}\cdot\left\langle\nu\middle|\boldsymbol{J}^\dagger(\boldsymbol{k})\middle|\nu_o\right\rangle\right|^2 D(E_\nu - E_{\nu_o} - \hbar\omega)\,.$$

If we suppose that $I_\mu(\omega)$ and the matrix element are slowly varying functions of ω at $\omega = \omega_{\nu\nu_o}$ we obtain

$$dw^{(\mu)}_{\nu_0\to\nu} = \frac{4\pi^2 I_\mu(\omega_{\nu\nu_0})}{\hbar^2\omega^2_{\nu\nu_0}c}\left|\hat{\boldsymbol{e}}_{k\mu}\cdot\left\langle\nu\middle|\boldsymbol{J}^\dagger(\boldsymbol{k})\middle|\nu_0\right\rangle\right|^2 d\Omega\ . \tag{4.21}$$

We remark that, for long times, this process only takes place if $\omega_{\nu\nu_o} = \omega_k$. Suppose now that, at time $t = 0$, the system is in the state ν. We wish to find $dw^{(\mu)}_{\nu\to\nu_o}$. The matrix element for this process is given in Eq. (4.16) where $n_{k\mu}$ is the photon occupation number in the state r of the radiation field. Now,

$$\hat{\boldsymbol{e}}_{k\mu}\cdot\langle\nu_0|\boldsymbol{J}(\boldsymbol{k})|\nu\rangle = \hat{\boldsymbol{e}}_{k\mu}\cdot\langle\nu|\boldsymbol{J}^\dagger(\boldsymbol{k})|\nu_0\rangle^* \tag{4.22}$$

so that the only difference between $dw^{(\mu)}_{\nu_o\to\nu}$ and $dw^{(\mu)}_{\nu\to\nu_o}$ is in the factor $n_{k\mu}+1$ to be compared with simply $n_{k\mu}$. This means, of course, that while absorption cannot occur if $n_{k\mu} = 0$, emission is always possible. We write

$$dw^{(\mu)}_{\nu\to\nu_0} = dw^{(\mu)}_{\nu_0\to\nu} + dw^{(\mu)}_{\nu\to\nu_0,s}\ .$$

Thus, $dw^{(\mu)}_{\nu\to\nu_0}$ equals the sum of a term proportional to the occupation number $n_{k\mu}$ of photons in the initial state of the radiation field identical to $dw^{(\mu)}_{\nu_0\to\nu}$ and $dw^{(\mu)}_{\nu\to\nu_0,s}$, the probability per unit time of spontaneous emission of radiation. From Eqs. (4.14) and (4.16) (the latter with $n_{k\mu} = 0$) we obtain

$$dw^{(\mu)}_{\nu\to\nu_0,s} = \frac{1}{t}\sum_{k\epsilon d\Omega}\frac{2\pi}{V\hbar\omega_k}|\hat{\boldsymbol{e}}_{k\mu}\cdot\langle\nu_0|\boldsymbol{J}(\boldsymbol{k})|\nu\rangle|^2\left|\int_0^t \exp[i(\omega_k - \omega_{\nu\nu_0})t']dt'\right|^2 .$$

Thus,

$$dw^{(\mu)}_{\nu\to\nu_0,s} = \frac{\omega_{\nu\nu_0}}{2\pi\hbar c^3}\,|\hat{\boldsymbol{e}}_{k\mu}\cdot\langle\nu_0|\boldsymbol{J}(\boldsymbol{k})|\nu\rangle|^2 d\Omega\ . \tag{4.23}$$

Now, the total transition probability for absorption of radiation in a transition $\nu_o \to \nu$ is

$$w_{\nu_o\to\nu} = \sum_\mu\int d\Omega\ \frac{4\pi^2 I_\mu(\omega_{\nu\nu_0})}{\hbar^2\omega^2_{\nu\nu_0}c}\left|\hat{\boldsymbol{e}}_{k\mu}\cdot\boldsymbol{J}^\dagger_{\nu\nu_o}(\boldsymbol{k})\right|^2 . \tag{4.24}$$

If $I_\mu(\omega)$ is the spectral density for an isotropic unpolarized distribution of radiation as that which exists in a cavity in thermal equilibrium we have

$$I_\mu(\omega) = \frac{cu(\omega)}{8\pi} \tag{4.25}$$

where u(ω) is the density of radiation per unit volume and per unit frequency range at ω. Then

$$w_{\nu_o\to\nu} = \frac{\pi u(\omega_{\nu\nu_o})}{2\hbar^2\omega_{\nu\nu_o}^2}\sum_\mu \int d\Omega\left|\hat{e}_{k\mu}\cdot \boldsymbol{J}_{\nu\nu_o}^\dagger(\boldsymbol{k})\right|^2. \tag{4.26}$$

Here, in order to save space we write $\boldsymbol{J}_{\nu\nu_0}^\dagger(\boldsymbol{k})$ for $\langle\nu|\boldsymbol{J}^\dagger(\boldsymbol{k})|\nu_0\rangle$. The Einstein B coefficient is defined by

$$w_{\nu_0\to\nu} = B_{\nu_o\nu}u(\omega_{\nu\nu_o}) \tag{4.27}$$

so that

$$B_{\nu_o\nu} = \frac{\pi}{2\hbar^2\omega_{\nu\nu_o}^2}\sum_\mu \int d\Omega|\hat{e}_{k\mu}\cdot \boldsymbol{J}_{\nu\nu_o}^\dagger(\boldsymbol{k})|^2 . \tag{4.28}$$

In the same way, the Einstein A coefficient is

$$A_{\nu\nu_0} = \sum_\mu \int \frac{dw_{\nu\to\nu_0,s}^{(\mu)}}{d\Omega}\, d\Omega = \frac{\omega_{\nu\nu_0}}{2\pi\hbar c^3}\sum_\mu \int d\Omega|\hat{e}_{k\mu}\cdot \boldsymbol{J}_{\nu_0\nu}(\boldsymbol{k})|^2 . \tag{4.29}$$

Comparing Eqs. (4.28) and (4.29) we find after making use of Eq. (4.22), that

$$\frac{A_{\nu\nu_o}}{B_{\nu_0\nu}} = \frac{\hbar\omega_{\nu\nu_o}^3}{\pi^2c^3}, \tag{4.30}$$

in agreement with the result derived by Einstein using the correspondence principle.

Consider now the matrix element

$$\hat{e}_{k\mu}\cdot \boldsymbol{J}_{\nu_0\nu}(\boldsymbol{k}) = \sum_i\left\{\frac{q_i}{m_i}\langle\nu_0|\hat{e}_{k\mu}\cdot \boldsymbol{p}_i\exp(-i\boldsymbol{k}\cdot\boldsymbol{r}_i)|\nu\rangle + ic\langle\nu_0|\hat{e}_{k\mu}\cdot(\boldsymbol{k}\times\boldsymbol{\mu}_i)\exp(-i\boldsymbol{k}\cdot\boldsymbol{r}_i)|\nu\rangle\right\} \tag{4.31}$$

and imagine that we are dealing with states $|\nu>$ and $|\nu_o>$ such that the particles have negligible probability of being found in a region of space outside a finite, fixed region of maximum diameter a. Let us, for simplicity take the origin of the coordinate system in the interior of the region where the particles are likely to be found. If this is not done, we simply multiply the matrix element by a constant phase factor $\exp(-i\boldsymbol{k}\cdot\boldsymbol{r}_0)$, where $\boldsymbol{r}_0$ is a position in the interior of the system. Now $|\boldsymbol{k}\cdot\boldsymbol{r}_i| \leq 2\pi|\boldsymbol{r}_i|/\lambda$ where λ is the wavelength of the radiation. If $(2\pi a/\lambda) \ll 1$, then we can approximate $\boldsymbol{J}(\boldsymbol{k})$ by disregarding the difference between $\exp(-i\boldsymbol{k}\cdot\boldsymbol{r}_i)$ and unity. This occurs for atoms emitting radiation in the visible region of the electromagnetic spectrum because $a \sim 10^{-8}$ cm and $\lambda \approx 6\times10^{-5}$ cm so that $|\boldsymbol{k}\cdot\boldsymbol{r}_i| \lesssim 10^{-3}$. For a single particle, we can write

$$\boldsymbol{J}(\boldsymbol{k}) = \frac{q}{m}\boldsymbol{p} + \sum_{\ell=1}^\infty \frac{(-ik)^\ell}{\ell!}\left\{\frac{q}{2m}(\boldsymbol{p}x_k^\ell + x_k^\ell\boldsymbol{p}) - \ell c\hat{\boldsymbol{k}}\times\boldsymbol{\mu}x_k^{\ell-1}\right\} \tag{4.32}$$

where

$$x_k = \frac{\boldsymbol{k}\cdot\boldsymbol{r}}{|\boldsymbol{k}|} = \hat{\boldsymbol{k}}\cdot\boldsymbol{r}\ . \tag{4.33}$$

Each term in the expansion (4.32) is $\sim 10^{-3}$ times smaller than the preceding one for the case just mentioned. We can write

$$\boldsymbol{J}(\boldsymbol{k}) = \sum_{\ell=0}^{\infty} k^{\ell} \boldsymbol{J}^{(\ell)}(\hat{\boldsymbol{k}})\ . \tag{4.34}$$

Here

$$\boldsymbol{J}^{(0)} = \sum_i \frac{q_i}{m_i}\boldsymbol{p}_i\ , \tag{4.35}$$

$$\boldsymbol{J}^{(1)}(\hat{\boldsymbol{k}}) = -i\sum_i\left\{\frac{q_i}{2m_i}\Big(\boldsymbol{p}_i\boldsymbol{r}_i\cdot\hat{\boldsymbol{k}} + \boldsymbol{r}_i\cdot\hat{\boldsymbol{k}}\boldsymbol{p}_i\Big) - c\hat{\boldsymbol{k}}\times\boldsymbol{\mu}_i\right\}\ . \tag{4.36}$$

The operators $J^{(\ell)}(\hat{\boldsymbol{k}})$ are defined in similar ways for $\ell \geq 2$. The operator $\boldsymbol{J}^{(0)}$ can be rewritten in the following form

$$\boldsymbol{J}^{(0)} = \frac{1}{i\hbar}[\boldsymbol{d}, H_M] \tag{4.37}$$

where

$$\boldsymbol{d} = \sum_i q_i\boldsymbol{r}_i \tag{4.38}$$

is the electric dipole moment operator.

The next term in the expansion is

$$k\hat{\boldsymbol{e}}_{k\mu}\cdot\boldsymbol{J}^{(1)}(\hat{\boldsymbol{k}}) = -ik\sum_i\left\{\frac{q_i}{m_i}\hat{\boldsymbol{k}}\cdot\boldsymbol{r}_i\boldsymbol{p}_i\cdot\hat{\boldsymbol{e}}_{k\mu} - c\Big(\hat{\boldsymbol{e}}_{k\mu}\times\hat{\boldsymbol{k}}\Big)\cdot\boldsymbol{\mu}_i\right\} \tag{4.39}$$

where we made use of the fact that $\hat{\boldsymbol{e}}_{k\mu}\cdot\boldsymbol{r}_i$ and $\hat{\boldsymbol{k}}\cdot\boldsymbol{p}_i$ commute. We can rewrite the tensor

$$\sum_i \frac{q_i}{m_i}\boldsymbol{r}_i\boldsymbol{p}_i = \sum_i \frac{q_i}{2m_i}(\boldsymbol{r}_i\boldsymbol{p}_i + \boldsymbol{p}_i\boldsymbol{r}_i) + \sum_i \frac{q_i}{2m_i}(\boldsymbol{r}_i\boldsymbol{p}_i - \boldsymbol{p}_i\boldsymbol{r}_i) \tag{4.40}$$

and use the commutation relation

$$[x_{i\alpha}x_{i\beta}, H_M] = \frac{i\hbar}{m_i}(x_{i\alpha}p_{i\beta} + p_{i\alpha}x_{i\beta})\ ,\ \alpha,\beta = x,y,z\ , \tag{4.41}$$

to give physical interpretations to the matrix elements of $J^{(1}(\hat{\boldsymbol{k}})$. For this purpose we define the electric quadrupole tensor operator $\boldsymbol{Q}$ whose components are

$$Q_{\alpha\beta} = \sum_i q_i\big(3x_{i\alpha}x_{i\beta} - r_i^2\delta_{\alpha\beta}\big)\ . \tag{4.42}$$

We find that

$$\left[\hat{\boldsymbol{k}} \cdot \boldsymbol{Q} \cdot \hat{\boldsymbol{e}}_{k\mu}, H_M\right] = 3i\hbar \sum_i \frac{q_i}{m_i} \hat{\boldsymbol{k}} \cdot (\boldsymbol{r}_i \boldsymbol{p}_i + \boldsymbol{p}_i \boldsymbol{r}_i) \cdot \hat{\boldsymbol{e}}_{k\mu} \; .$$

Thus,

$$k\hat{\boldsymbol{e}}_{k\mu} \cdot \boldsymbol{J}^{(1)}(\hat{\boldsymbol{k}}) = -\frac{k}{6\hbar}\left[\hat{\boldsymbol{k}} \cdot \boldsymbol{Q} \cdot \hat{\boldsymbol{e}}_{k\mu}, H_M\right] + ick\left(\hat{\boldsymbol{e}}_{k\mu} \times \hat{\boldsymbol{k}}\right) \cdot \sum_i \left(\frac{q_i}{2m_i c}\boldsymbol{r}_i \times \boldsymbol{p}_i + \boldsymbol{\mu}_i\right) . \tag{4.43}$$

The sum in the second term is, simply, the total magnetic moment operator $\boldsymbol{\mu}$ of the system. Thus we write

$$k\hat{\boldsymbol{e}}_{k\mu} \cdot \boldsymbol{J}^{(1)}(\hat{\boldsymbol{k}}) = -\frac{k}{6\hbar}\left[\hat{\boldsymbol{k}} \cdot \boldsymbol{Q} \cdot \hat{\boldsymbol{e}}_{k\mu}, H_M\right] + ick\left(\hat{\boldsymbol{e}}_{k\mu} \times \hat{\boldsymbol{k}}\right) \cdot \boldsymbol{\mu} \; . \tag{4.44}$$

In our studies we usually need

$$\langle \nu_o | \boldsymbol{J}^{(0)} | \nu \rangle = \frac{1}{i\hbar} \langle \nu_o | [\boldsymbol{d}, H_M] | \nu \rangle = -i\omega_{\nu\nu_o} \langle \nu_o | \boldsymbol{d} \, | \nu \rangle \; , \tag{4.45}$$

and

$$\left\langle \nu_o \middle| k\hat{\boldsymbol{e}}_{k\mu} \cdot \boldsymbol{J}^{(1)}(\hat{\boldsymbol{k}}) \middle| \nu \right\rangle = -\frac{1}{6}\, k\omega_{\nu\nu_0} \left\langle \nu_0 \middle| \hat{\boldsymbol{k}} \cdot \boldsymbol{Q} \cdot \hat{\boldsymbol{e}}_{k\mu} \middle| \nu \right\rangle + ikc \left\langle \nu_o \middle| (\hat{\boldsymbol{e}}_{k\mu} \times \hat{\boldsymbol{k}}) \cdot \boldsymbol{\mu} \middle| \nu \right\rangle . \tag{4.46}$$

The ratio of the orders of magnitude of (4.46) and (4.45) when neither vanishes is $(ka)^{-1}$. For atoms this is about 10^{-3} as we have seen. The approximation in which only Eq. (4.45) is kept is called the electric dipole approximation. The terms in (4.46) give rise to the electric quadrupole radiation and the magnetic dipole radiation. In the dipole approximation the results (4.23) and (4.29) become

$$dw^{(\mu)}_{\nu \to \nu_0, s} = \frac{\omega^3_{\nu\nu_0}}{2\pi\hbar c^3} \left| \hat{\boldsymbol{e}}_{k\mu} \cdot \boldsymbol{d}_{\nu_0\nu} \right|^2 d\Omega \tag{4.47}$$

and

$$A_{\nu\nu_0} = \sum_\mu \int dw^{(\mu)}_{\nu\to\nu_0,s} = \frac{\omega^3_{\nu\nu_0}}{2\pi\hbar c^3} \int \sin\theta d\theta d\varphi |\boldsymbol{d}_{\nu_0\nu}|^2 \sin^2\theta = \frac{4\omega^3_{\nu\nu_0}}{3\hbar c^3} |\boldsymbol{d}_{\nu_0\nu}|^2 \; , \tag{4.48}$$

respectively. To carry out the derivation of Eq. (4.48) we took $\boldsymbol{d}_{\nu_0\nu}$ as polar axis; only the polarization in the plane of $\boldsymbol{d}_{\nu_0\nu}$ and $\boldsymbol{k}$ gives a contribution to the total transition probability. The power radiated in a transition from $\nu \to \nu_o$ is

$$P_{\nu\to\nu_o,s} = \frac{4\omega^4_{\nu\nu_0}}{3c^3} |\boldsymbol{d}_{\nu_0\nu}|^2 \; . \tag{4.49}$$

We complete this section with a discussion of the selection rules appropriate to electric dipole, electric quadrupole and magnetic dipole transitions. The components of $\boldsymbol{d}$ can be written in terms of their irreducible components defined by

$$d_{\pm 1} = \mp \frac{1}{\sqrt{2}} (d_x \pm i d_y) \, , \; d_o = d_z \; . \tag{4.50}$$

The Wigner-Eckart[7] theorem

$$\langle n'j'm'|d_\kappa|njm\rangle = (2j'+1)^{-1/2}\langle n'j' \parallel \boldsymbol{d} \parallel nj\rangle\langle j1;j'm'|j1;m\kappa\rangle \ , \qquad (4.51)$$

allows us to express the matrix elements of $d_\kappa(\kappa = 0, \pm 1)$ in terms of a reduced matrix element $\langle n'j' \parallel \boldsymbol{d} \parallel nj\rangle$ and a Clebsch-Gordan coefficient $\langle j1;j'm'|j1;m\kappa\rangle$. The former is independent of the "angular" quantum numbers κ, m and m', so that the dependence on κ, m and m' is entirely in the Clebsch-Gordan coefficient.

This shows that the matrix elements of the d_κ vanish unless $|j'-j| = 0, 1$ and $\kappa = m' - m$. Thus $\Delta m = 0, \pm 1$. If the Hamiltonian of the system is invariant under the operation of parity $P(\boldsymbol{r} \to \boldsymbol{r}' = -\boldsymbol{r})$, its eigenvectors can be chosen so that they have definite parity, i.e., $P|\nu\rangle = \pm|\nu\rangle$. However,

$$P^{-1}\boldsymbol{d}P = P^\dagger \boldsymbol{d}P = -\boldsymbol{d} \qquad (4.52)$$

and

$$\langle\nu_o|\boldsymbol{d}|\nu\rangle = -\langle\nu_o|P^\dagger\boldsymbol{d}P|\nu\rangle = -\langle P\nu_o|\boldsymbol{d}|P\nu\rangle \qquad (4.53)$$

which vanishes unless the parities of ν and ν_o are different. Thus the selection rules for dipole transitions are:

$$\Delta j = 0, \pm 1; \ \Delta m = 0, \pm 1, \ \text{ change in parity} \ . \qquad (4.54)$$

If we consider a spinless particle in a central field, $j = \ell$. The transition $\Delta\ell = 0$ is now forbidden because the parity of $Y_\ell^m(\theta,\varphi)$ is $(-1)^\ell$. There is an additional restriction for electromagnetic transitions. Transitions between two states with $j = 0$ are forbidden to all orders of $\boldsymbol{k}$. We say that $j = 0 \to j = 0$ transitions are strictly forbidden. To prove this we start from the observation that if $|\nu\rangle$ and $|\nu_o\rangle$ have zero total angular momentum they are invariant under all rotations, i.e.,

$$\boldsymbol{J}|\nu>=|\nu>, \boldsymbol{J}|\nu_o>=|\nu_0> . \qquad (4.55)$$

where $\boldsymbol{J}$ is the total angular momentum of the system. Consider now a rotation by 180° about $\boldsymbol{k}$; then, if the symbol R denotes this operation, $P_R^\dagger \boldsymbol{k}\cdot\boldsymbol{r}_i P_R = \boldsymbol{k}\cdot\boldsymbol{r}_i$, $P_R^\dagger \hat{\mathbf{e}}_{k\mu}\cdot\boldsymbol{p}_i P_R = -\hat{\mathbf{e}}_{k\mu}\cdot\boldsymbol{p}_i$ and $P_R^\dagger(\hat{\mathbf{e}}_{k\mu}\times\boldsymbol{k})\cdot\boldsymbol{\mu}_i P_R = -(\hat{\mathbf{e}}_{k\mu}\times\boldsymbol{k})\cdot\boldsymbol{\mu}_i$. The last equation follows from the fact that $(\hat{\mathbf{e}}_{k\mu}\times\boldsymbol{k})\cdot\boldsymbol{\mu}_i$ is the component of $\boldsymbol{\mu}_i$ along a vector perpendicular to both $\hat{\mathbf{e}}_{k\mu}$ and $\boldsymbol{k}$. If we set $\boldsymbol{k}$ along the z-axis and $\hat{\mathbf{e}}_{k\mu}$ along $\hat{\boldsymbol{x}}$, then $(\hat{\mathbf{e}}_{k\boldsymbol{\mu}}\times\boldsymbol{k})\cdot\boldsymbol{\mu}_i = -k\mu_{iy}$. But under P_R, μ_y transforms as $\hbar L_y = zp_x - xp_z$ and our statement is proved. Thus, from Eq. (4.8) we have

$$P_R^\dagger\hat{\mathbf{e}}_{k\mu}\cdot\boldsymbol{J}(\boldsymbol{k})P_R = -\hat{\mathbf{e}}_{k\mu}\cdot\boldsymbol{J}(\boldsymbol{k}) . \qquad (4.56)$$

Therefore

$$\langle\nu_o|P_R^\dagger\hat{\mathbf{e}}_{k\mu}\cdot\boldsymbol{J}(\boldsymbol{k})P_R|\nu\rangle = -\langle\nu_o|\hat{\mathbf{e}}_{k\mu}\cdot\boldsymbol{J}(\boldsymbol{k})|\nu\rangle$$

But from Eqs. (4.55), $P_R|\nu\rangle = |\nu\rangle$ and $P_R|\nu_o\rangle = |\nu_o\rangle$ so that

$$\langle\nu_o|\hat{\mathbf{e}}_{k\mu}\cdot\boldsymbol{J}(\boldsymbol{k})|\nu\rangle = 0 . \qquad (4.57)$$

The selection rules appropriate to electric quadrupole transitions are those for a second rank symmetric tensor. By virtue of the Wigner-Eckart theorem, the matrix elements are proportional to

$$\langle jk, j'm'|jk; m\kappa\rangle$$

where $k = 2$, $\kappa = 0, \pm1, \pm2$. Thus $m' = m + \kappa$ and $|j' - j| = 0, 1, 2$ and, hence, $\Delta j = 0, \pm1, \pm2$, $\Delta m = 0, \pm1, \pm2$. If $[H_M, P] = 0$ there is no change in parity between the initial and final states.

For a single, spinless particle $\Delta\ell = 0, \pm2$ and $\Delta m = 0, \pm1, \pm2$. As before transitions between states with $j = 0$ are strictly forbidden.

Chapter 5

The Scattering of Light by Matter

Our purpose in this section is to give a quantum mechanical derivation of the cross section for light scattering. The process which interests us is one in which the states r_o and r differ in that r contains one less photon in mode $(\boldsymbol{k}, \hat{\mathbf{e}})$ than r_o and one more in the state $(\boldsymbol{k}', \hat{\mathbf{e}}')$, while at the same time the scattering system experiences a transition from ν_o to ν. In order to provide sufficient generality, instead of expanding the vector potential $\boldsymbol{A}(\boldsymbol{r})$ in terms of plane polarized components, we consider expansions with arbitrary polarization. These are expressed, for each $\boldsymbol{k}$, in terms of two complex-valued unit vectors $\hat{\mathbf{e}}_k$ orthogonal to $\boldsymbol{k}$ and to each other. For example, for circularly polarized waves propagating along the $\hat{\boldsymbol{z}}$-axis with respect to the right-handed triad $(\hat{\boldsymbol{x}}, \hat{\boldsymbol{y}}, \hat{\boldsymbol{z}})$ we use

$$\hat{\mathbf{e}}_{\pm} = \frac{1}{\sqrt{2}}(\hat{\boldsymbol{x}} \pm i\hat{\boldsymbol{y}}) \ .$$

The upper (lower) sign corresponds to left- (right-) handed circularly polarized radiation. Equivalently we speak of positive (negative) helicity. The vectors $\hat{\mathbf{e}}_+, \hat{\mathbf{e}}_-$ and $\hat{\boldsymbol{z}}$ form a complete set for vectors in 3-space. They are orthonormal in the sense that $\hat{\mathbf{e}}^*_{\pm} \cdot \hat{\mathbf{e}}_{\pm} = 1$, $\hat{\mathbf{e}}^*_{\pm} \cdot \hat{\boldsymbol{z}} = 0$ and $\hat{\mathbf{e}}^*_{\pm} \cdot \hat{\mathbf{e}}_{\mp} = 0$. For details see Appendix I.

To calculate the probability for this process to occur, we use Eq. (4.13) with H'_{II} in first order and H'_I in second order. Let n and n$'$ be the numbers of photons in states $(\boldsymbol{k}, \hat{\mathbf{e}})$ and $(\boldsymbol{k}', \hat{\mathbf{e}}')$ in r_o, respectively. The matrix element

$$\langle r\nu | H'_{II}(t') | r_o \nu_o \rangle$$

consists of two equal contributions both of which are represented by the diagram in Fig. 5.1. Set $\omega_k = \omega$, $\omega_{k'} = \omega'$. From Eq. (4.6),

$$\langle r\nu | H'_{II}(t') | r_o \nu_o \rangle = \frac{2\pi\hbar}{V} \left[\frac{n(n'+1)}{\omega\omega'} \right]^{1/2} \hat{e}'^* \cdot \hat{e} \exp[i\,(\omega_{\nu\nu_0} + \omega' - \omega)\,t'] \left\langle \nu \left| \sum_i \frac{q_i^2}{m_i} \exp[i(\boldsymbol{k} - \boldsymbol{k}') \cdot \boldsymbol{r}_i] \right| \nu_o \right\rangle . \tag{5.1}$$

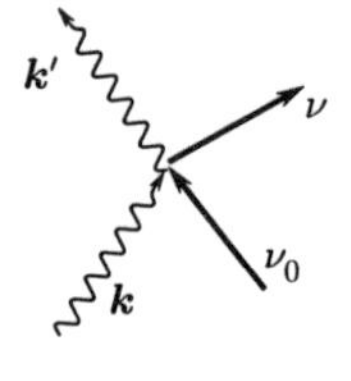

Figure 5.1 Diagram illustrating the contribution of A^2 to the scattering matrix element in first order perturbation theory.

The matrix element

$$\langle r\nu|H_I'(t_2)H_I'(t_1)|r_o\nu_o\rangle = \sum_{r'\nu'}\langle r\nu|H_I'(t_2)|r'\nu'\rangle\langle r'\nu'|H_I'(t_1)|r_o\nu_o\rangle \tag{5.2}$$

involves summations over all intermediate states. There are, however, only two possibilities for r'. Since $|r_o\rangle = |\ldots n, n', \ldots\rangle$ and $|r\rangle = |\ldots n-1, n'+1\ldots\rangle$, r' can only be $|\ldots n-1, n'\ldots\rangle$ or $|\ldots n, n'+1\ldots\rangle$. These two possibiliites are represented by the diagrams in Fig. 5.2.

We obtain

$$< r\nu|H_I(t_2)H_I(t_1)|r_0\nu_0 >= \sum_{\nu'}\frac{2\pi\hbar}{V}\left[\frac{n(n'+1)}{\omega\omega'}\right]^{1/2}$$
$$\cdot\Bigg(\hat{e}'^{*}\cdot \boldsymbol{J}_{\nu\nu'}(\boldsymbol{k}')\hat{e}\cdot\boldsymbol{J}^{\dagger}_{\nu'\nu_0}(\boldsymbol{k})\exp\Big(i(\omega_{\nu\nu'}+\omega')t_2\Big)\exp\Big(i(\omega_{\nu'\nu_0}-\omega)t_1\Big)$$
$$+\hat{e}\cdot\boldsymbol{J}^{\dagger}_{\nu\nu'}(\boldsymbol{k})\hat{e}'^{*}\cdot\boldsymbol{J}_{\nu'\nu_0}(\boldsymbol{k}')\exp\Big(i(\omega_{\nu\nu'}-\omega)t_2\Big)\exp\Big(i(\omega_{\nu'\nu_0}+\omega')t_1\Big)\Bigg).$$

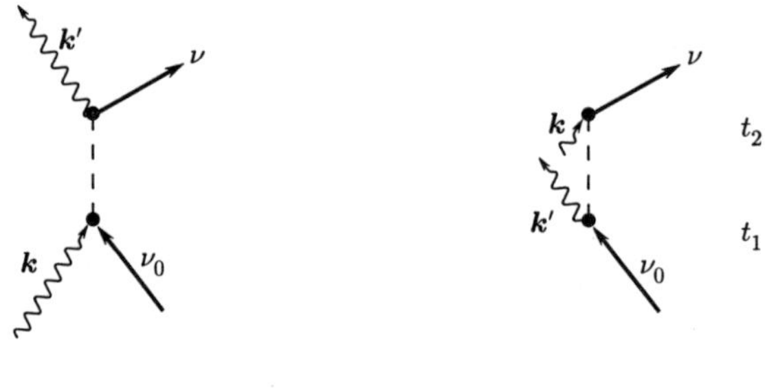

Figure 5.2 Diagrams showing the contributions of $\boldsymbol{A}$ to the scattering matrix element in second order perturbation theory.

We recall, however, that what we need is

$$-\frac{1}{\hbar^2}\int_0^t dt_2 \int_0^{t_2} dt_1 < r\nu\ |H_I(t_2)H_I(t_1)|r_0\nu_0 >= \sum_{\nu'}$$
$$\frac{2\pi}{V\hbar}\left[\frac{n(n'+1)}{\omega\omega'}\right]^{1/2}\Bigg[\hat{e}'^{*}\cdot \boldsymbol{J}_{\nu\nu'}(\boldsymbol{k}')\hat{e}\cdot \boldsymbol{J}^{\dagger}_{\nu'\nu_0}(\boldsymbol{k})\left(\frac{\exp\Big(i(\omega_{\nu\nu_0}+\omega'-\omega)t\Big)-1}{(\omega_{\nu'\nu_0}-\omega)(\omega_{\nu\nu_0}+\omega'-\omega)}\right.$$
$$\left.+\frac{\exp\Big(i(\omega_{\nu\nu'}+\omega')t\Big)-1}{(\omega_{\nu'\nu_0}-\omega)(\omega_{\nu\nu'}+\omega')}\right)$$
$$+\hat{e}\cdot \boldsymbol{J}^{\dagger}_{\nu\nu'}(\boldsymbol{k})\hat{e}'^{*}\cdot \boldsymbol{J}_{\nu'\nu_0}(\boldsymbol{k}')\left(\frac{\exp\Big(i(\omega_{\nu\nu_0}+\omega'-\omega)t\Big)-1}{(\omega_{\nu'\nu_0}+\omega')(\omega_{\nu\nu_0}+\omega'-\omega)}\right.$$
$$\left.-\frac{\exp\Big(i(\omega_{\nu\nu'}-\omega)t\Big)-1}{(\omega_{\nu'\nu_0}+\omega')(\omega_{\nu\nu'}-\omega)}\right)\Bigg]. \tag{5.3}$$

In the process we are considering the conservation of energy requires that

$$\omega_{\nu\nu_o}-\omega+\omega' = 0\,. \tag{5.4}$$

The quantities

$$\frac{\exp(i(\omega_{\nu\nu'}+\omega')t)-1}{(\omega_{\nu\nu'}+\omega')(\omega_{\nu'\nu_0}-\omega)} \quad\text{and}\quad \frac{\exp(i(\omega_{\nu\nu'}-\omega)t)}{(\omega_{\nu\nu'}-\omega)(\omega_{\nu'\nu_0}+\omega')}\,.$$

will be neglected because they become large only when $\omega' = \omega_{\nu'\nu}$ and $\omega = \omega_{\nu\nu'}$, respectively. In conjunction with Eq. (5.4) these require that $\omega = \omega_{\nu'\nu_0}$ and $\omega' = \omega_{\nu_0\nu'}$, respectively. The first corresponds to absorption of the incident photon and excitation from ν_o to ν' while the second represents emission of a photon of frequency ω' in a transition from ν_o to ν'. Often one assumes that ν_o is the ground state so that the second process is impossible. The first requires that ω be a possible absorption frequency but we usually consider the scattering of incident light to which the system is transparent. In evaluating the probability that a transition occurs, the terms neglected yield small periodic contributions for sufficiently long times. Then, the expression (5.3) becomes

$$\frac{2\pi}{\hbar V}\left[\frac{n(n'+1)}{\omega\omega'}\right]^{1/2}\sum_{\nu'}\left\{\frac{\hat{e}'^{*}\cdot \boldsymbol{J}_{\nu\nu'}(\boldsymbol{k}')\hat{e}\cdot \boldsymbol{J}^{\dagger}_{\nu'\nu_0}(\boldsymbol{k})}{\omega_{\nu'\nu_0}-\omega}+\frac{\hat{e}\cdot \boldsymbol{J}^{\dagger}_{\nu\nu'}(\boldsymbol{k})\hat{e}'^{*}\cdot \boldsymbol{J}_{\nu'\nu_0}(\boldsymbol{k}')}{\omega_{\nu'\nu_0}+\omega'}\right\}$$
$$\cdot\frac{e^{i(\omega_{\nu\nu_0}+\omega'-\omega)t}-1}{\omega_{\nu\nu_0}+\omega'-\omega}$$

We can now write

$$-\frac{i}{\hbar}\int_0^t \langle r\nu|H'_{II}(t')|r_0\nu_0\rangle dt' - \frac{1}{\hbar^2}\int_0^t dt_2\int_0^{t_2} dt_1\langle r\nu|H'_I(t_2)H'_I(t_1)|r_0\nu_0\rangle$$
$$= \frac{2\pi}{\hbar V}\left[\frac{n(n'+1)}{\omega\omega'}\right]^{1/2} M_{\nu\nu_0}\,\frac{\exp(i(\omega_{\nu\nu_o}+\omega'-\omega)t-1}{\omega_{\nu\nu_o}+\omega'-\omega} \tag{5.5}$$

where

$$M_{\nu\nu_0} = \sum_{\nu'} \left(\frac{\hat{e}'^* \cdot \boldsymbol{J}_{\nu\nu'}(\boldsymbol{k}')\hat{e} \cdot \boldsymbol{J}^{\dagger}_{\nu'\nu_0}(\boldsymbol{k})}{\omega_{\nu'\nu_0} - \omega} + \frac{\hat{e} \cdot \boldsymbol{J}^{\dagger}_{\nu\nu'}(\boldsymbol{k})\hat{e}'^* \cdot \boldsymbol{J}_{\nu'\nu_0}(\boldsymbol{k}')}{\omega_{\nu'\nu} + \omega} \right) - \hbar \hat{e}'^* \cdot \hat{e} \left\langle \nu \middle| \sum_i \frac{q_i^2}{m_i} \exp(i(\boldsymbol{k} - \boldsymbol{k}') \cdot \boldsymbol{r}_i) \middle| \nu_0 \right\rangle . \tag{5.6}$$

Thus

$$P_{r_o\nu_o \to r\nu}(t) = \frac{4\pi^2}{\hbar^2 V^2} \frac{n(n'+1)}{\omega\omega'} |M_{\nu\nu_0}|^2 D(E_\nu - E_{\nu_o} + \hbar\omega' - \hbar\omega) \tag{5.7}$$

and, for large t, we find

$$w_{r_o\nu_o \to r\nu} = \frac{1}{t} P_{r_o\nu_o \to r\nu}(t) = \frac{8\pi^3}{\hbar^2 V^2} \frac{n(n'+1)}{\omega\omega'} |M_{\nu\nu_0}|^2 \delta(\omega_{\nu\nu_0} + \omega' - \omega) . \tag{5.8}$$

The cross section for scattering of a photon into the solid angle $d\Omega'$ is

$$\begin{aligned} d\sigma &= \sum_{k' \epsilon d\Omega'} \frac{8\pi^3}{\hbar^2 V^2} \frac{n(n'+1)}{\omega\omega'} |M_{\nu\nu_0}|^2 \delta(\omega_{\nu\nu_0} + \omega' - \omega) \frac{V}{nc} \\ &= d\Omega' \frac{V}{8\pi^3} \int \frac{\omega'^2 d\omega'}{c^3} \frac{8\pi^3}{\hbar^2 V^2} \frac{n(n'+1)}{\omega\omega'} |M_{\nu\nu_0}|^2 \delta(\omega_{\nu\nu_0} + \omega' - \omega) \frac{V}{nc} , \end{aligned}$$

or

$$d\sigma = (\omega'/\omega\hbar^2 c^4) |M_{\nu\nu_0}|^2 (n'+1) \, d\Omega' , \tag{5.9}$$

where it is understood that ω' is given by Eq. (5.4).

If we consider scattering of visible light by molecules whose dimensions are of the order of a few Angstrom units and molecular excitations having energies much less than $\hbar\omega$, we can the use the electric dipole approximation and replace $\boldsymbol{J}(\boldsymbol{k})$ in Eq. (5.6) by $\boldsymbol{J}^{(0)}$ as given in Eq. (4.35). The last term in Eq. (5.6) becomes

$$-\hbar \hat{e}'^* \cdot \hat{e} \delta_{\nu\nu_0} \sum_i \frac{q_i^2}{m_i} . \tag{5.10}$$

We saw in Eq. (4.45) that

$$\boldsymbol{J}^{(0)}_{\nu'\nu} = i\omega_{\nu'\nu} \boldsymbol{d}_{\nu'\nu} . \tag{5.11}$$

The operator $\boldsymbol{J}^{(0)}$ does not depend on $\boldsymbol{k}$ and is Hermitian. Equation (5.6), in the electric dipole approximation reads

$$M_{\nu\nu_0} = \sum_{\nu'} \left(\frac{\hat{e}'^* \cdot \boldsymbol{d}_{\nu\nu'} \hat{e} \cdot \boldsymbol{d}_{\nu'\nu_0}}{\omega - \omega_{\nu'\nu_0}} - \frac{\hat{e} \cdot \boldsymbol{d}_{\nu\nu'} \hat{e}'^* \cdot \boldsymbol{d}_{\nu'\nu_0}}{\omega_{\nu'\nu} + \omega} \right) \omega_{\nu\nu'} \omega_{\nu'\nu_0} - \hbar \hat{e}'^* \cdot \hat{e} \delta_{\nu\nu_0} \sum_i \frac{q_i^2}{m_i} . \tag{5.12}$$

We note that $M_{\nu\nu_0}$ is invariant under the substitutions $\hat{e} \leftrightarrow \hat{e}'^*$, $\omega \leftrightarrow -\omega'$. This is shown directly from Eq. (5.12) or from consideration of time reversal symmetry. (It is also true in general with the additional substitution $\boldsymbol{k} \leftrightarrow -\boldsymbol{k}'$.)

To simplify Eq. (5.12) we consider the sum

$$\sum_{\nu'} (\omega_{\nu\nu'}\omega_{\nu'\nu_0} + \omega\omega') \left(\frac{\hat{e}'^* \cdot \boldsymbol{d}_{\nu\nu'}\ \hat{e} \cdot \boldsymbol{d}_{\nu'\nu_0}}{\omega - \omega_{\nu'\nu_0}} - \frac{\hat{e} \cdot \boldsymbol{d}_{\nu\nu'}\ \hat{e}'^* \cdot \boldsymbol{d}_{\nu'\nu_0}}{\omega' + \omega_{\nu'\nu_0}} \right) \tag{5.13}$$

and note that

$$\omega_{\nu\nu'}\omega_{\nu'\nu_0} + \omega\omega' = (\omega - \omega_{\nu'\nu_0})(\omega' + \omega_{\nu'\nu_0}). \tag{5.14}$$

Thus, the expression (5.13) is identical to

$$\begin{aligned}
&\sum_{\nu'} \left[(\omega' + \omega_{\nu'\nu_0})(\hat{e}'^* \cdot \boldsymbol{d}_{\nu\nu'}\ \hat{e} \cdot \boldsymbol{d}_{\nu'\nu_0}) - (\omega - \omega_{\nu'\nu_0})(\hat{e} \cdot \boldsymbol{d}_{\nu\nu'}\ \hat{e}'^* \cdot \boldsymbol{d}_{\nu'\nu_0})\right] \\
&= \omega \sum_{\nu'} (\hat{e}'^* \cdot \boldsymbol{d}_{\nu\nu'}\ \hat{e} \cdot \boldsymbol{d}_{\nu'\nu_0} - \hat{e} \cdot \boldsymbol{d}_{\nu\nu'}\ \hat{e}'^* \cdot \boldsymbol{d}_{\nu'\nu_0}) \\
&\quad + \sum_{\nu'} (\omega_{\nu'\nu_0}\ \hat{e} \cdot \boldsymbol{d}_{\nu\nu'}\ \hat{e}'^* \cdot \boldsymbol{d}_{\nu'\nu_0} - \omega_{\nu\nu'}\ \hat{e}'^* \cdot \boldsymbol{d}_{\nu\nu'}\ \hat{e} \cdot \boldsymbol{d}_{\nu'\nu_0})
\end{aligned}$$

where we replaced $\omega' + \omega_{\nu'\nu_0}$ with $\omega - \omega_{\nu\nu'}$. The first sum vanishes exactly since it equals

$$\omega \hat{e}'^* \cdot (\boldsymbol{dd})_{\nu\nu_0} \cdot \hat{e} - \omega \hat{e} \cdot (\boldsymbol{dd})_{\nu\nu_0} \cdot \hat{e}'^* \equiv 0\,.$$

To evaluate the second sum we use Eq. (5.11) to obtain

$$\begin{aligned}
&\sum_{\nu'} (\omega_{\nu'\nu_0} \hat{e} \cdot \boldsymbol{d}_{\nu\nu'} \hat{e}'^* \cdot \boldsymbol{d}_{\nu'\nu_0} - \omega_{\nu\nu'} \hat{e}'^* \cdot \boldsymbol{d}_{\nu\nu'} \hat{e} \cdot \boldsymbol{d}_{\nu'\nu_0}) \\
&= \frac{1}{i} \sum_{\nu'} (\hat{e} \cdot \boldsymbol{d}_{\nu\nu'} \hat{e}'^* \cdot \boldsymbol{J}^{(0)}_{\nu'\nu_0} - \hat{e}'^* \cdot \boldsymbol{J}^{(0)}_{\nu\nu'} \hat{e} \cdot \boldsymbol{d}_{\nu'\nu_0}) \\
&= \frac{1}{i} \left[\hat{e} \cdot \boldsymbol{d}, \hat{e}'^* \cdot \boldsymbol{J}^{(0)}\right]_{\nu\nu_0}.
\end{aligned}$$

But the commutator of $\hat{e} \cdot \boldsymbol{d}$ and $\boldsymbol{e}'^* \cdot \boldsymbol{J}^{(0)}$ is

$$[\hat{e} \cdot \boldsymbol{d}, \hat{e}'^* \cdot \boldsymbol{J}^{(0)}] = i\hbar \hat{e}'^* \cdot \hat{e} \sum_i \frac{q^2}{m_i}\,.$$

Combining these results we deduce that

$$\begin{aligned}
&\sum_{\nu'} (\omega_{\nu\nu'}\omega_{\nu'\nu_0} + \omega\omega') \left(\frac{\hat{e}'^* \cdot \boldsymbol{d}_{\nu\nu'}\ \hat{e} \cdot \boldsymbol{d}_{\nu'\nu_0}}{\omega - \omega_{\nu'\nu_0}} - \frac{\hat{e} \cdot \boldsymbol{d}_{\nu\nu'}\ \hat{e}'^* \cdot \boldsymbol{d}_{\nu'\nu_0}}{\omega' + \omega_{\nu'\nu_0}} \right) \\
&= \hbar \hat{e}'^* \cdot \hat{e} \delta_{\nu\nu_0} \sum_i \frac{q_i^2}{m_i}\,.
\end{aligned} \tag{5.15}$$

Thus, Eqs. (5.12) and (5.15) give

$$\mathrm{M}_{\nu\nu_0} = -\omega\omega' \left(\frac{\hat{e}'^* \cdot \boldsymbol{d}_{\nu\nu'}\ \hat{e} \cdot \boldsymbol{d}_{\nu'\nu_0}}{\omega - \omega_{\nu'\nu_0}} - \frac{\hat{e} \cdot \boldsymbol{d}_{\nu\nu'}\ \hat{e}'^* \cdot \boldsymbol{d}_{\nu'\nu_0}}{\omega' + \omega_{\nu'\nu_0}} \right). \tag{5.16}$$

The cross section is, thus, expressed in the form

$$\frac{d\sigma}{d\Omega'} = \left(\frac{{\omega'}^3\omega}{\hbar^2 c^4}\right)\left|\sum_{\nu'}\left(\frac{\hat{e}\cdot \boldsymbol{d}_{\nu\nu'}\ \hat{e}'^*\cdot \boldsymbol{d}_{\nu'\nu_0}}{\omega' + \omega_{\nu'\nu_0}} - \frac{\hat{e}'^*\cdot \boldsymbol{d}_{\nu\nu'}\ \hat{e}\cdot \boldsymbol{d}_{\nu'\nu_0}}{\omega - \omega_{\nu'\nu_0}}\right)\right|^2 (n'+1), \tag{5.17}$$

If the scattering occurs in a medium of index of refraction n, the speed of light c must be replaced with c/n.

We investigate now the form of the cross section in Eq. (5.17) when ω is sufficiently large so that most of the scattering amplitude originates from matrix elements of the electric dipole moment between states of energies much less than $\hbar\omega$. To be true the sum over intermediate states extends to higher energy states. The high frequency limit, to be derived below, is valid when the matrix elements $\boldsymbol{d}_{\nu'\nu_0}$ and $\boldsymbol{d}_{\nu\nu'}$ decrease rapidly with increasing $E_{\nu'}$. We also suppose that $\omega_{\nu\nu_0} \ll \omega$.

Under these conditions we separate the sum over ν' into two parts, the first, which we suppose to be predominant, extends over states such that $E_{\nu'} - E_\nu \ll \hbar\omega$ while the second includes all other intermediate states. These may become important when $\omega \sim \omega_{\nu'\nu_0}$, *i.e.*, under resonant conditions. We now obtain an expansion of the cross section (5.17) in powers of ω^{-1}. We replace $\omega' + \omega_{\nu'\nu_0}$ with $\omega - \omega_{\nu\nu'}$ and expand to obtain

$$\frac{d\sigma}{d\Omega'} = \frac{{\omega'}^3}{\omega\hbar^2 c^4}(n'+1)\left|\sum_{\nu'}\left(\left(1 + \frac{\omega_{\nu\nu'}}{\omega} + \frac{{\omega_{\nu\nu'}}^2}{\omega^2} + \cdots\right)\hat{e}\cdot \boldsymbol{d}_{\nu\nu'}\ \hat{e}'^*\cdot \boldsymbol{d}_{\nu'\nu_0}\right.\right.$$
$$\left.\left. - \left(1 + \frac{\omega_{\nu'\nu_0}}{\omega} + \frac{{\omega_{\nu'\nu_0}}^2}{\omega^2} + \cdots\right)\hat{e}'^*\cdot \boldsymbol{d}_{\nu\nu'}\ \hat{e}\cdot \boldsymbol{d}_{\nu'\nu_0}\right)\right|^2 .$$

Now

$$\sum_{\nu'}\left(\hat{e}\cdot \boldsymbol{d}_{\nu\nu'}\ \hat{e}'^*\cdot \boldsymbol{d}_{\nu'\nu_0} - \hat{e}'^*\cdot \boldsymbol{d}_{\nu\nu'}\ \hat{e}\cdot \boldsymbol{d}_{\nu'\nu_0}\right) = 0$$

if we extend the sum over ν' over all states of the scattering system. The next term in the expansion of the amplitude is proportional to the matrix element of the double commutator $[\hat{e}'^*\cdot \boldsymbol{d}\,,\ [\hat{e}\cdot \boldsymbol{d}, H_0]]$ between $|\nu>$ and $|\nu_0>$. We obtain

$$\frac{d\sigma}{d\Omega'} = \frac{n'+1}{(\hbar c)^4}\left(\frac{\omega'}{\omega}\right)^3 \left|\langle\nu|[\hat{e}'^*\cdot \boldsymbol{d}\,,\ [\hat{e}\cdot \boldsymbol{d}, H_0]]\nu_0\rangle + O\left(\frac{1}{\omega}\right)\right|^2 . \tag{5.18}$$

For elastic scattering from a non-degenerate ground state

$$\frac{d\sigma}{d\Omega'} = (n'+1)|\hat{e}'^*\cdot\hat{e}|^2\left(\sum_i \frac{q_i^2}{m_i c^2}\right)^2 . \tag{5.19}$$

Equation (5.19) is valid only in the high frequency limit. For example, consider an electron harmonically bound to a center of force by an attractive force characterized by a resonant frequency ω_0. The steady state displacement of the electron under the action of an electric field $\boldsymbol{E}$ varying harmonically with angular frequency ω is $(e\boldsymbol{E}/m)(\omega^2 - \omega_0^2)^{-1}$

The electric field at $\boldsymbol{r}$ due to the induced dipole moment $(e^2\boldsymbol{E}/m)(\omega_0^2 - \omega^2)^{-1}$ is, according to Eqs. (1.2) and (1.3),

$$\boldsymbol{E}' = \frac{e^2 E}{mc^2 r} \frac{\omega^2}{\omega^2 - \omega_0^2} (\hat{\mathbf{e}} \times \hat{\boldsymbol{n}}) \times \hat{\boldsymbol{n}} \tag{5.20}$$

where (e^2/mc^2) is the classical radius of the electron. If the incident field is unpolarized we can regard it as having two components of equal intensity, one polarized in the plane of scattering while the other is polarized at right angles to that plane. For the first $\hat{\mathbf{e}}' \cdot \hat{\mathbf{e}} = \cos\theta$ while for the second $\hat{\mathbf{e}}' \cdot \hat{\mathbf{e}} = 1$. Thus, the differential scattering cross section is

$$\frac{d\sigma}{d\Omega'} = \frac{1}{2} \left(\frac{e^2}{mc^2} \right)^2 \left(\frac{\omega^2}{\omega^2 - \omega_0^2} \right)^2 (1 + \cos^2\theta) \, . \tag{5.21}$$

Upon integration over the angular variables we obtain the total scattering cross section

$$\sigma = \frac{8\pi}{3} \left(\frac{e^2}{mc^2} \right)^2 \left(\frac{\omega^2}{\omega^2 - \omega_0^2} \right)^2 \, . \tag{5.22}$$

This is the Thomson scattering cross section. In the high frequency limit ($\omega \gg \omega_0$) Eq. (5.21) yields a result identical to that given in Eq. (5.19) for the particular situation of the simple model considered.

Chapter 6

Light Scattering in Solids

The dipole approximation was proved above for molecules but it can be shown that it is also valid when the scattering system is a crystal even though its diameter is much larger than the wavelength of the incident light. We shall establish the validity of the dipole approximation when the diameter of the primitive cell is small compared to $2\pi|\boldsymbol{k}|^{-1}$. To prove this, we first review the Bloch theorem.

We consider a crystal whose primitive translations are the three non-coplanar vectors $\boldsymbol{a}_1, \boldsymbol{a}_2, \boldsymbol{a}_3$. The Hamiltonian H_M is invariant under translations by

$$\boldsymbol{n} = n_1\boldsymbol{a}_1 + n_2\boldsymbol{a}_2 + n_3\boldsymbol{a}_3 \ , \quad n_1, n_2, n_3 = 0, \pm 1, \pm 2, \dots \ . \tag{6.1}$$

A many-body wave function $\psi(\boldsymbol{r}_1, \boldsymbol{r}_2, \dots \boldsymbol{r}_N)$ transforms under the translation operator $T(\boldsymbol{n})$ according to

$$T(\boldsymbol{n})\psi(\boldsymbol{r}_1, \boldsymbol{r}_2, \dots \boldsymbol{r}_N) = \psi(\boldsymbol{r}_1 + \boldsymbol{n}, \boldsymbol{r}_2 + \boldsymbol{n}, \dots \boldsymbol{r}_N + \boldsymbol{n}) \ . \tag{6.2}$$

Let ψ be a stationary state of H_M with eigenvalue E, *i.e.*, it satisfies the Schrödinger equation

$$H_M\psi = E\psi \ . \tag{6.3}$$

Now, since

$$T(\boldsymbol{n})H_M T^{-1}(\boldsymbol{n}) = H_M \ , \tag{6.4}$$

$T(\boldsymbol{n})\psi$ is also an eigenstate of H_M with the same eigenvalue E. The operator $T(\boldsymbol{n})$ is unitary. Thus, for every $\boldsymbol{n}$ a Hermitian operator $S(\boldsymbol{n})$ exists such that

$$T(\boldsymbol{n}) = \exp(iS(\boldsymbol{n})) \ . \tag{6.5}$$

Since any two translation operators commute,

$$T(\boldsymbol{n})T(\boldsymbol{n}') = T(\boldsymbol{n}')T(\boldsymbol{n}) = T(\boldsymbol{n} + \boldsymbol{n}') \ , \tag{6.6}$$

the corresponding Hermitian operators $S(\boldsymbol{n})$ possess the properties

$$[S(\boldsymbol{n}), S(\boldsymbol{n}')] = 0 \tag{6.7}$$

and

$$S(\boldsymbol{n}) + S(\boldsymbol{n}') = S(\boldsymbol{n} + \boldsymbol{n}') \ . \tag{6.8}$$

Furthermore, since according to Eq. (6.4), $T(\boldsymbol{n})$ and H_M commute so do $S(\boldsymbol{n})$ and H_M and hence, the sets of operators $\{S(\boldsymbol{n})\}$, H_M possess a common set of eigenvectors which form a basis for the Hilbert space of the system described by H_M. Let ψ be an element of this basis and $s(\boldsymbol{n})$ the eigenvalue of $S(\boldsymbol{n})$. Since $T(0)$ is the identity operation, for all states $s(0) = 1$. In addition

$$s(\boldsymbol{n} + \boldsymbol{n}') = s(\boldsymbol{n}) + s(\boldsymbol{n}') \tag{6.9}$$

by virtue of Eq. (6.8). The only linear function mapping vectors in three-dimensional space into the real numbers is a projection operator. Therefore, for each state ψ of the basis described above, a vector $\boldsymbol{q}$ exists such that

$$s(\boldsymbol{n}) = \boldsymbol{q} \cdot \boldsymbol{n} \ . \tag{6.10}$$

This leads immediately to the Bloch theorem expressed in the form

$$T(\boldsymbol{n})\psi(\boldsymbol{r}_1, \boldsymbol{r}_2 \ldots \boldsymbol{r}_N) = \exp(i\boldsymbol{q} \cdot \boldsymbol{n})\psi(\boldsymbol{r}_1, \boldsymbol{r}_2 \ldots \boldsymbol{r}_N) \ . \tag{6.11}$$

Such states are called Bloch states. To each we can associate a vector $\boldsymbol{q}$. However, there is a redundancy if we consider all such vectors. In fact, two vectors, $\boldsymbol{q}$ and $\boldsymbol{q}'$, such that $(\boldsymbol{q}' - \boldsymbol{q}) \cdot \boldsymbol{n}$ is an integral multiple of 2π for all $\boldsymbol{n}$, give rise to the same transformation under translations. The vectors $\boldsymbol{G}$ such that, for all $\boldsymbol{n}$, $\boldsymbol{G} \cdot \boldsymbol{n}$ is an integral multiple of 2π form a lattice whose primitive translations are the vectors $\boldsymbol{b}_1, \boldsymbol{b}_2, \boldsymbol{b}_3$ defined by the nine conditions $\boldsymbol{a}_i \cdot \boldsymbol{b}_j = 2\pi\delta_{ij}$. These vectors form the so-called reciprocal lattice of the crystal and any two vectors $\boldsymbol{q}$ and $\boldsymbol{q}'$ which differ by a vector of the reciprocal lattice are to be regarded as equivalent. We therefore restrict $\boldsymbol{q}$ to the region defined by the inequalities

$$-\pi < \boldsymbol{q} \cdot \boldsymbol{a}_i \leq \pi \ , \quad i = 1, 2, 3 \ . \tag{6.12}$$

This region in $\boldsymbol{q}$-space is called the fundamental Brillouin zone of the crystal.

When considering a finite, rather than an infinite, crystal possessing $N = N_0^3$ primitive cells defined by $n_i = 0, 1, 2, \ldots N_0 - 1$ $(i = 1, 2, 3)$ we use periodic boundary conditions in which all quantities (measurable or not) have periods $N_0\boldsymbol{a}_1, N_0\boldsymbol{a}_2, N_0\boldsymbol{a}_3$ (*i.e.*, $\psi(\{\boldsymbol{r}_i + N_0\boldsymbol{n}\}) = \psi(\{\boldsymbol{r}_i\})$ for all $\boldsymbol{n}$). These boundary conditions, even though unphysical, are appropriate when considering properties originating in the bulk of the material rather than on its surface. The validity of these restrictions is ensured for extensive properties when N is sufficiently large for the effects of the surface atoms to be negligible. The ratio of the number of surface to bulk atoms is of the order of $N_0^2/N_0^3 = N_0^{-1} = N^{-1/3}$. The periodic boundary conditions restrict the possible values of $\boldsymbol{q}$. In fact, for all $\boldsymbol{n}$, $N_0\boldsymbol{q} \cdot \boldsymbol{n}$ must be an integral multiple of 2π. Thus $N_0\boldsymbol{q} \cdot \boldsymbol{a}_i = 2\pi\ell_i$ $(i = 1, 2, 3; \ell_i = 0, \pm1, \pm2, \ldots)$. Hence

$$\boldsymbol{q} = \frac{1}{N_0}(\ell_1\boldsymbol{b}_1 + \ell_2\boldsymbol{b}_2 + \ell_3\boldsymbol{b}_3) \ . \tag{6.13}$$

Because of the redundancy of values of $\boldsymbol{q}$ (see Eq. (6.12)) it is enough to select the integers ℓ_i so that

$$-\tfrac{1}{2}\, N_0 < \ell_i \leq \tfrac{1}{2}\, N_0 \ . \tag{6.14}$$

Clearly there are N_0 values of ℓ_i verifying this inequality and, hence, $N_0^3 = N$ values of $\boldsymbol{q}$ in the fundamental Brillouin zone. Another, seldom used, choice of the values of ℓ_i is $0, 1, 2, \ldots N_0 - 1$.

We are now in a position to prove a wave vector conservation law for Raman scattering by crystals. For this purpose we study the matrix elements of $\boldsymbol{J}(\boldsymbol{k}')$ (see Eq. (4.8)). We write

$$\boldsymbol{J}_{\nu\nu'}(\boldsymbol{k}') = \int d\boldsymbol{r}_1 d\boldsymbol{r}_2 \ldots d\boldsymbol{r}_N \psi_\nu^*(\boldsymbol{r}_1, \boldsymbol{r}_2, \ldots \boldsymbol{r}_N) \boldsymbol{J}(\boldsymbol{k}') \psi_{\nu'}(\boldsymbol{r}_1, \boldsymbol{r}_2, \ldots \boldsymbol{r}_N) \ . \tag{6.15}$$

This quantity is invariant if all $\boldsymbol{r}_i$ are translated by $\boldsymbol{n}$, *i.e.*,

$$\boldsymbol{J}_{\nu\nu'}(\boldsymbol{k}') =$$
$$\int d\boldsymbol{r}_1 d\boldsymbol{r}_2 \ldots d\boldsymbol{r}_N \psi_\nu^*(\boldsymbol{r}_1+\boldsymbol{n}, \ldots \boldsymbol{r}_N+\boldsymbol{n}) T(\boldsymbol{n}) \boldsymbol{J}(\boldsymbol{k}') T^{-1}(\boldsymbol{n}) \psi_{\nu'}(\boldsymbol{r}_1+\boldsymbol{n}, \ldots \boldsymbol{r}_N+\boldsymbol{n}) \ .$$

But, by virtue of Eq. (4.8)

$$T(\boldsymbol{n}) \boldsymbol{J}(\boldsymbol{k}') T^{-1}(\boldsymbol{n}) = \exp(-i\boldsymbol{k}' \cdot \boldsymbol{n}) \boldsymbol{J}(\boldsymbol{k}') \ . \tag{6.16}$$

We now apply Bloch's theorem. We let $\boldsymbol{q}$, $\boldsymbol{q}'$ and $\boldsymbol{q}_0$ be the wave vectors associated with the states ν, ν' and ν_0, respectively. Then

$$\boldsymbol{J}_{\nu\nu'}(\boldsymbol{k}') = \exp[-i(\boldsymbol{q} + \boldsymbol{k}' - \boldsymbol{q}') \cdot \boldsymbol{n}] \boldsymbol{J}_{\nu\nu'}(\boldsymbol{k}') \ , \tag{6.17}$$

and $\boldsymbol{J}_{\nu\nu'}(\boldsymbol{k}')$ vanishes unless

$$\boldsymbol{q} + \boldsymbol{k}' - \boldsymbol{q}' = 0 \ , \tag{6.18}$$

or its equivalent, any vector of the reciprocal lattice. The first term in Eq. (5.6) vanishes except for those terms in which ν' is such that $\boldsymbol{q} + \boldsymbol{k}' - \boldsymbol{q}' = 0$ and $\boldsymbol{q}' - (\boldsymbol{k} + \boldsymbol{q}_0) = 0$. Thus, only states ν for which

$$\boldsymbol{q} + \boldsymbol{k}' = \boldsymbol{q}_0 + \boldsymbol{k} \tag{6.19}$$

give a non-zero contribution. Identical conclusions are reached for the second and third terms in $M_{\nu\nu_0}$. We emphasize that when a restriction such as that in Eq. (6.19) is written it is implied that the equality holds up to the addition of an arbitrary vector of the reciprocal lattice.

We can now prove the validity of the dipole approximation for crystals. We remember that $\boldsymbol{J}(\boldsymbol{k})$ is an operator involving the sum of single particle terms of the form $\exp(-i\boldsymbol{k} \cdot \boldsymbol{r})\boldsymbol{p}$ so that the matrix elements of $\boldsymbol{J}(\boldsymbol{k})$ for many-particle states can be expressed as linear combinations of integrals of the form

$$\int d\boldsymbol{r} \psi_f^* \exp(-i\boldsymbol{k} \cdot \boldsymbol{r}) \boldsymbol{p} \psi_i$$

where ψ_i and ψ_f are one-particle Bloch functions. This can be rewritten as the sum over all primitive cells $\boldsymbol{n}$

$$\sum_n \int_{\Omega_n} \psi_f^* \exp(-i\boldsymbol{k}\cdot\boldsymbol{r})\boldsymbol{p}\psi_i d\boldsymbol{r}$$

where the integral extends over the volume Ω_n of the primitive cell translated by $\boldsymbol{n}$ with respect to Ω_0, the primitive cell around the origin of the coordinate system. In each of these integrals we set $\boldsymbol{r} = \boldsymbol{r}' + \boldsymbol{n}$ and use Bloch's theorem to obtain

$$\begin{aligned}\sum_n \int_{\Omega_n} & \psi_f^*(\boldsymbol{r}'+\boldsymbol{n}) \exp(-i\boldsymbol{k}\cdot(\boldsymbol{r}'+\boldsymbol{n}))\boldsymbol{p}\psi_i(\boldsymbol{r}'+\boldsymbol{n})d\boldsymbol{r} \\ &= \sum_n \exp[i(\boldsymbol{q}_i - \boldsymbol{k} - \boldsymbol{q}_f)\cdot\boldsymbol{n}] \int_{\Omega_0} \psi_f^*(\boldsymbol{r}')e^{-i\boldsymbol{k}\cdot\boldsymbol{r}'}\boldsymbol{p}'\psi_i(\boldsymbol{r}')d\boldsymbol{r}'\end{aligned}$$

We note that the last integral is independent of $\boldsymbol{n}$ and, that throughout the volume Ω_0, $|\boldsymbol{k}\cdot\boldsymbol{r}'| \sim ka$ where a is a length of the order of magnitude of the lattice parameter. Thus, whenever $ka \ll 1$ we can set $\exp(-i\boldsymbol{k}'\cdot\boldsymbol{r}) = 1$ and the dipole approximation is valid.

Let us suppose the scattering system has a center of inversion. In such a case the stationary states can be chosen so that they have a definite parity, *i.e.*, they are even or odd under inversion ($\boldsymbol{r} \to \boldsymbol{r}' = -\boldsymbol{r}$) with respect to the center of symmetry. The matrix element

$$\langle\nu|\boldsymbol{d}|\nu'\rangle \neq 0$$

only if the states ν and ν' have different parity because $\boldsymbol{d}$ is odd under inversion. Thus, the transition $\nu_0 \to \nu$ is possible in Raman scattering only if intermediate states ν' exist such that $\langle\nu|\boldsymbol{d}|\nu'\rangle$ and $\langle\nu'|\boldsymbol{d}|\nu_0\rangle$ are different from zero. This is possible only if ν_0 and ν have the same parity. We contrast this to the fact that (in the dipole approximation) in ordinary absorption and emission of radiation a transition from ν_0 to ν involving a single photon is only possible if ν_0 and ν have different parities.

It is customary to say that a transition taking place between two states ν_0 and ν with emission or absorption of radiation is infra-red active. The word infra-red is used because the large majority of such investigations in molecules and many in solids involve infra-red radiation. A transition $\nu_0 \to \nu$ accompanied by inelastic scattering of a photon is said to be Raman active. Thus, within the dipole approximation, in systems having inversion symmetry, Raman and infra-red activity are mutually exclusive. Hence Raman and absorption spectroscopy are complementary tools in the study of such systems. Before studying the Raman spectra of crystals which do not possess a center of inversion we review the theory of group representations.

Chapter 7

Application of Symmetry Principles to the Raman Spectroscopy of Solids: Vibrational Modes.

Not all selection rules are as simply obtained as those we studied above. For many systems it becomes necessary to use the theory of group representations.[8] We introduce the subject by means of an example. The structure of α-quartz is similar to that of tellurium which is shown in Fig. 7.1. The primitive cell contains three SiO_2 units. Inspection of the structure reveals that the following operations leave the crystal in a position which cannot be distinguished from its initial configuration:

the identity E,
rotations by 120° and 240° about the trigonal or optic axis,
rotations by 180° about axes lying in a plane perpendicular to the trigonal axis and forming angles of 120° with each other.

To discuss these operations further we shall introduce a convenient notation. We take a Cartesian coordinate system with its z-axis parallel to the trigonal axis. The rotation by $120^o = 2\pi/3$ is designated by C_3 and that by 240° by C_3^2. These two operations form a class in a technical use of the word which we need not discuss yet. This class will be denoted by $2C_3$. The number 2 designates the number of elements in the class and C_3 is a typical such element. Provisionally we can think of the elements in a class as being "similar" to one another. The x-axis of our coordinate system is taken along one of the two-fold axis described above. We denote that particular operation of rotation by $180^o = 2\pi/2$ by the symbol C_2. The two others are designated by C_2' and C_2''. The three operations C_2, C_2' and C_2'' form a class which we denote by $3C_2$. The six operations $\{E, C_3, C_3^2, C_2, C_2', C_2''\}$

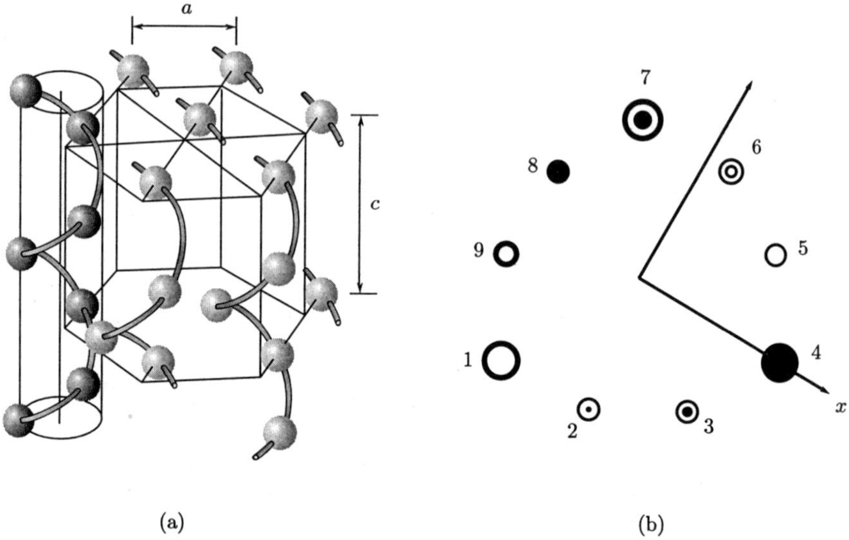

Figure 7.1 The spiral structure of α–quartz about the z axis is shown in part (a) where each solid ball represents a SiO_2 unit. The projection of the atoms onto the basal plane is shown in part (b). The Si atoms (1, 4 and 7) occupy positions at levels 0, $c/3$ and $2c/3$ respectively while the oxygen atoms (9, 5, 3, 8, 6 and 2) occupy positions at levels $c/9$, $2c/9$, $4c/9$, $5c/9$, $7c/9$ and $8c/9$ respectively. The coordinate axes shown are those used in this work. Note that the space group of α–quartz is D_3^4 or the enantiomorphous D_3^6. The figures show the right–handed structure. (S. Rodriguez and A. K. Ramdas, Inelastic Light Scattering in Crystals in "Highlights of Condensed Matter Theory". *Proceedings of the International School of Physics, "Enrico Fermi"* Course LXXXIX, F. Bassani, F. Fumi, and M. P. Tosi, editors (North Holland, 1985) pp 369-420).

are the rotations about the center of gravity of an equilateral triangle which leave it invariant. The successive application of any two of these six operations is, itself, an operation of the set. Further, each element of the set has an inverse in the set. In fact, C_2, C_2' and C_2'' are their own inverses and C_3 and C_3^2 are inverses of each other. Thus the set $\{E, C_3, C_3, C_2, C_2', C_2''\}$ forms a group which, traditionally, is denoted by the symbol D_3. We now consider the effect of these operations on the vector $\boldsymbol{r} = (x, y, z)$. The action of C_3 transforms $\boldsymbol{r}$ into $\boldsymbol{r}' = (x', y', z')$,

$$\boldsymbol{r}' = C_3 \boldsymbol{r} \ , \tag{7.1}$$

where

$$x' = -\frac{1}{2}x - \frac{\sqrt{3}}{2}y \ , \qquad y' = \frac{\sqrt{3}}{2}\,x - \frac{1}{2}\,y \ , \qquad z' = z \ . \tag{7.2}$$

The operation C_3 can be represented by the matrix

$$\Gamma(C_3) = \begin{pmatrix} -\frac{1}{2} & -\frac{\sqrt{3}}{2} & 0 \\ \frac{\sqrt{3}}{2} & -\frac{1}{2} & 0 \\ 0 & 0 & 1 \end{pmatrix} \ . \tag{7.3}$$

The matrix $\Gamma(C_3^2)$ can be found from $\boldsymbol{r} \to \boldsymbol{r}' = C_3(C_3\boldsymbol{r})$ so that $\Gamma(C_3^2) = \Gamma^2(C_3)$ and

$$\Gamma(C_3^2) = \begin{pmatrix} -\frac{1}{2} & \frac{\sqrt{3}}{2} & 0 \\ -\frac{\sqrt{3}}{2} & -\frac{1}{2} & 0 \\ 0 & 0 & 1 \end{pmatrix} . \tag{7.4}$$

The operation C_2 gives rise to the transformation

$$\boldsymbol{r}' = C_2\boldsymbol{r} , \tag{7.5}$$

or

$$\begin{aligned} x' &= x \\ y' &= -y \\ z' &= -z \end{aligned} \tag{7.6}$$

which is represented by the matrix

$$\Gamma(C_2) = \begin{pmatrix} 1 & 0 & 0 \\ 0 & -1 & 0 \\ 0 & 0 & -1 \end{pmatrix} . \tag{7.7}$$

For each operation of the group $D_3 = \{E, 2C_3, 3C_2\}$ we can construct a 3×3 matrix. The result of two successive transformations on $\boldsymbol{r}$ gives rise to a third whose matrix is obtained by multiplying the corresponding matrices. The set of matrices $\Gamma(A)$ where A ranges over the elements of D_3 is said to be a representation of the group D_3. The representation which we have just constructed is three-dimensional. An interesting feature of this representation is that it can be decomposed into simpler components. In fact, under all operations of D_3, vectors parallel to the z-axis transform into vectors parallel to the z-axis and vectors in the (x, y) plane transform into vectors in the (x, y) plane. This result can be stated in a slightly different way as follows. The three-dimensional vector $\boldsymbol{r}$ can be written in a unique way as the sum of $\boldsymbol{r}_\perp$ and $\boldsymbol{r}_\parallel$,

$$\boldsymbol{r} = \boldsymbol{r}_\perp + \boldsymbol{r}_\parallel = (\boldsymbol{r} - \boldsymbol{r} \cdot \boldsymbol{z}\hat{\boldsymbol{z}}) + \boldsymbol{r} \cdot \hat{\boldsymbol{z}}\hat{\boldsymbol{z}} , \tag{7.8}$$

where $\boldsymbol{r}_\perp$ lies in the (x, y) plane and $\boldsymbol{r}_\parallel$ is parallel to $\hat{\boldsymbol{z}}$. We say that the subspaces $\{\boldsymbol{r}_\perp\}$ and $\{\boldsymbol{r}_\parallel\}$ are invariant under the operations of D_3 and that the three-dimensional representation $\Gamma(A)$ is reducible. The subspace of vectors parallel to $\hat{\boldsymbol{z}}$ generate the representation of one-dimensional matrices

$$\Gamma(E) = 1 , \quad \Gamma(C_3) = \Gamma(C_3^2) = 1 , \quad \Gamma(C_2) = \Gamma(C_2') = \Gamma(C_2'') = -1 \tag{7.9}$$

while the vectors $\boldsymbol{r}_\perp$ generate the representation

$$\Gamma(E)=\begin{pmatrix}1&0\\0&1\end{pmatrix},\ \Gamma(C_3)=\begin{pmatrix}-\frac{1}{2}&-\frac{\sqrt{3}}{2}\\ \frac{\sqrt{3}}{2}&-\frac{1}{2}\end{pmatrix},\ \Gamma(C_3^2)=\begin{pmatrix}-\frac{1}{2}&\frac{\sqrt{3}}{2}\\ -\frac{\sqrt{3}}{2}&-\frac{1}{2}\end{pmatrix},$$
$$\Gamma(C_2)=\begin{pmatrix}1&0\\0&-1\end{pmatrix},\ \Gamma(C_2')=\begin{pmatrix}-\frac{1}{2}&-\frac{\sqrt{3}}{2}\\ -\frac{\sqrt{3}}{2}&\frac{1}{2}\end{pmatrix},\ \Gamma(C_2'')=\begin{pmatrix}-\frac{1}{2}&\frac{\sqrt{3}}{2}\\ \frac{\sqrt{3}}{2}&\frac{1}{2}\end{pmatrix}. \tag{7.10}$$

In the theory of group representations one proves that a representation of a group is characterized by the traces of its elements. The set of such numbers is called the character of a representation. Two elements of a group, A and B are said to belong to the same equivalence class if an element X of the group exists such that $B = X^{-1}AX$. Then, $\Gamma(B) = \Gamma(X^{-1}AX) = \Gamma(X^{-1})\Gamma(A)\Gamma(X) = \Gamma^{-1}(X)\Gamma(A)\Gamma(X)$ and $Tr\Gamma(B) = Tr\Gamma(A)$. Hence the matrices in a representation belonging to elements in the same class have the same trace. It is now a simple matter for the reader to verify that C_2, C_2' and C_2'' belong to the same class. So are C_3 and C_3^2 in a class containing two elements. The identity element is always in a class by itself.

A representation such that the space on which its elements act cannot be decomposed into invariant subspaces is said to be irreducible. We saw that the three-dimensional representation generated by the components of a vector $\boldsymbol{r}$ is reducible into the representations given in Eqs. (7.9) and (7.10). These are, however, irreducible. Two representation $\Gamma(A)$ and $\Gamma'(A)$ related by a similarity transformation

$$\Gamma'(A) = S^{-1}\Gamma(A)S \tag{7.11}$$

where S is a non-singular matrix are said to be equivalent and have the same character. Clearly two equivalent representations are obtained one from the other by a transformation of the basis vectors of the space in which they act. They are geometrically and physically the same object. Two theorems will now be stated without proof:

(1) The number of inequivalent irreducible representations of a group containing a finite number of elements is equal to the number of its equivalence classes.

(2) In a group containing h operations and c classes the dimensionalities $\ell_1, \ell_2, \ldots \ell_c$ of the inequivalent irreducible representations obey the restriction

$$\ell_1^2 + \ell_2^2 + \ldots + \ell_c^2 = h\,. \tag{7.12}$$

From this we deduce that D_3 has three inequivalent irreducible representations of which two are one-dimensional and the third is two-dimensional. They are labeled by the symbols Γ_1, Γ_2, and Γ_3, respectively; Γ_1 is the totally symmetric representation which associates the number one to every element of the group; Γ_2 is the representation of Eqs. (7.9) and Γ_3 that of Eqs. (7.10). It is customary to exhibit these characters in the form of a table. (see Table 7.1)

Table 7.1 *Character table for D_3*

D_3	E	$2C_3$	$3C_2$	Basis functions
Γ_1	1	1	1	$z^2; (x^2+y^2)$
Γ_2	1	1	-1	z
Γ_3	2	-1	0	$x, y; (x^2-y^2), -2xy; yz, -zx$

The last column of Table 7.1 gives functions which generate the different irreducible representations. Here x, y and z are components of a vector field transforming under the operations of rotation as the components x, y, z of a polar vector $\boldsymbol{r}$. Clearly both z^2 and (x^2+y^2) remain invariant under the action of all elements of the group. Thus they generate Γ_1. The quantity z remains the same under E and $2C_3$ but changes sign under $3C_2$; it generates Γ_2. To generate the two-dimensional representation Γ_3 we require two functions. Clearly x and y generate this representation. A little algebra convinces one that so do the other two pairs of functions listed in Table 7.1. The order and signs associated with these functions have been selected so that they generate the representation Γ_3 in exactly the same form as is generated by x and y, *i.e.*, in the form given by Eqs. (7.10). Other orders and signs are possible, however, but the representations thus generated, even though, equivalent to that generated by x and y appear in a different fashion. The reason for making this arrangement is that it allows the use of a theorem in the theory of group representations which we will quote without proof. The rows of the matrices in a representation are characterized by the functions in the basis generating it. If these functions are orthogonal to one another, the resulting matrices are unitary since the transformations of the group preserve inner products. Inspection of Eqs. (7.10) reveals that this representation is unitary. The vectors corresponding to x and y are orthogonal. Thus, the functions $(x^2-y^2)/2$ and $-xy$ are orthogonal and so are yz and $-zx$. The first row of the matrices in Eqs. (7.10) are said to belong to x and the second to y. Therefore we say that $(x^2-y^2)/2$ belongs to the first row of Γ_3 and $-xy$ to the second. The functions $x, y, z, \ldots$ may be regarded as quantum mechanical states. Associated with every energy eigenvalue there is an irreducible representation of the group. For a system whose Hamiltonian has symmetry D_3 the eigenvalues are either nondegenerate belonging to Γ_1 or Γ_2 or they are doubly degenerate belonging to Γ_3. The theorem we mentioned above can now be stated: The inner product of two basis functions is zero unless they belong to the same irreducible representation; furthermore, their inner product is zero for any two functions belonging to different rows of a unitary irreducible representation. This theorem says then that $\langle x|z\rangle = 0$ and $\langle x|y\rangle = 0$. In the same way, while $\langle x|\frac{1}{2}(x^2-y^2)\rangle \neq 0$, $\langle x|-xy\rangle = 0$.

We are now in a position to apply these ideas to a specific case. We study first the selection rules for absorption of radiation by a system such as α-quartz whose symmetry is D_3. A transition from ψ_0 to ψ_f occurs if $\langle\psi_f|\boldsymbol{d}|\psi_0\rangle \neq 0$. Let us suppose first that ψ_0 is the ground state with symmetry Γ_1. The components d_x, d_y, d_z of the dipole moment operator behave as the components X, Y, Z of a vector field. The quantities $d_x|\psi_0\rangle$ and $d_y|\psi_0\rangle$ belong to Γ_3 while $d_z|\psi_0\rangle$ belongs to Γ_2. If an excited state whose symmetry is Γ_2 is designated by Z while X and Y are states corresponding to a doubly degenerate excited state of symmetry Γ_3 we find that, in general

$$\langle X|d_x|\psi_0\rangle = \langle Y|d_y|\psi_0\rangle \neq 0 \tag{7.13}$$

and

$$\langle Z|d_z|\psi_0\rangle \neq 0 \ . \tag{7.14}$$

Other matrix elements vanish. Thus, states of symmetries Γ_2 and Γ_3 are infra-red active while those of symmetry Γ_1 are not. Further, states of symmetry Γ_2 are excited only by radiation polarized parallel to the z-axis (optic axis) and states of symmetry Γ_3 by radiation polarized perpendicular to this axis.

We consider now the polarizability tensor whose components are

$$\langle \nu|\alpha_{ij}|\nu_0\rangle = \sum_{\nu'} \left(-\frac{\langle \nu|d_i|\nu'\rangle\langle \nu'|d_j|\nu_0\rangle}{\omega - \omega_{\nu'\nu_0}} + \frac{\langle \nu'|d_i|\nu_0\rangle\langle \nu|d_j|\nu'\rangle}{\omega + \omega_{\nu'\nu}} \right) . \tag{7.15}$$

We introduce a new operator $\boldsymbol{b}$ whose quantum-mechanical equation of motion is

$$\left(i\frac{d}{dt} + \omega \right) \boldsymbol{b} = \frac{1}{\hbar}[\boldsymbol{b}, H_M] + \omega \boldsymbol{b} = \boldsymbol{d} \tag{7.16}$$

Taking matrix elements of both sides of Eq. (7.16) we find

$$\langle \nu|\boldsymbol{b}|\nu'\rangle = \frac{\langle \nu|\boldsymbol{d}|\nu'\rangle}{\omega + \omega_{\nu'\nu}} \tag{7.17}$$

Substituting this result in Eq. (7.15) and using the completeness of the states $|\nu\rangle$ we obtain

$$\langle \nu|\alpha_{ij}|\nu_0\rangle = \langle \nu|b_j d_i - d_i b_j|\nu_0\rangle \ . \tag{7.18}$$

Since both $\boldsymbol{b}$ and $\boldsymbol{d}$ transform like polar vectors, the components of α_{ij} transform under orthogonal transformations as a second rank tensor. Any second rank tensor can be rewritten in the form

$$\alpha_{ij} = \tfrac{1}{2} \left(\alpha_{ij} + \alpha_{ji} - \tfrac{2}{3}\, \delta_{ij} Tr \boldsymbol{\alpha} \right) + \tfrac{1}{2}\, (\alpha_{ij} - \alpha_{ji}) + \tfrac{1}{3}\, \delta_{ij} Tr \boldsymbol{\alpha}. \tag{7.19}$$

The first term,

$$\alpha_{ij}^{(S)} = \tfrac{1}{2} \left(\alpha_{ij} + \alpha_{ji} - \tfrac{2}{3}\, \delta_{ij} Tr \boldsymbol{\alpha} \right) \tag{7.20}$$

is a traceless symmetric tensor, the second

$$\alpha_{ij}^{(A)} = \tfrac{1}{2}\, (\alpha_{ij} - \alpha_{ji}) \tag{7.21}$$

is antisymmetric, and the last

$$\alpha_{ij}^{(D)} = \tfrac{1}{3}\, \delta_{ij} Tr \alpha \tag{7.22}$$

is a constant tensor, *i.e.*, a multiple of the unit tensor. In the absence of an external magnetic field or of internal modes of magnetization, $\alpha_{ij}^{(A)} = 0$. For the time being we can demonstrate this using a macroscopic argument. We must have

$$\alpha_{ij}(\boldsymbol{B}) = \alpha_{ji}(-\boldsymbol{B}) \tag{7.23}$$

or

$$\alpha_{ij}(\boldsymbol{M}) = \alpha_{ji}(-\boldsymbol{M}) \tag{7.24}$$

where $\boldsymbol{B}$ is the magnetic induction and $\boldsymbol{M}$ the magnetization. Clearly, if $\boldsymbol{B}$ and $\boldsymbol{M}$ vanish, the tensor α_{ij} is symmetric.

To demonstrate the way in which we apply the theory of groups we remember that the scattering cross section is proportional to $|\langle\nu|\hat{e}'^{*}\cdot\boldsymbol{\alpha}\cdot\hat{e}|\nu_0\rangle|^2$ where $\hat{e}$ and $\hat{e}'$ are the directions of polarization of the incident and scattered beams, respectively. We shall consider nonmagnetic excitations so that $\boldsymbol{\alpha}$ is symmetric. We consider the case of α-quartz with the $\boldsymbol{z}$-axis along the optic axis and the $\boldsymbol{x}$-axis along one of the two-fold axes of rotational symmetry. The α_{xx} components behaves under rotations in the same way as x^2. Thus $\frac{1}{2}\left(\alpha_{xx}+\alpha_{yy}\right)$ and α_{zz} are invariant under all operations of the group, *i.e.*, they belong to Γ_1; $\frac{1}{2}\left(\alpha_{xx}-\alpha_{yy}\right)$ and $-\alpha_{xy}$ is a pair of elements belonging to Γ_3, another pair being α_{yz} and $-\alpha_{zx}$. The representation generated by the six components of $\boldsymbol{\alpha}$ is reducible into Γ_1 and Γ_3 each appearing twice. We write symbolically

$$\Gamma_\alpha = 2\Gamma_1 + 2\Gamma_3\ . \tag{7.24}$$

Thus, if the ground state has symmetry Γ_1, only states of symmetries Γ_1 and Γ_3 can be final states in first order Raman scattering. We summarize these results in Table 7.2.

Table 7.2 *Selection rules for electric dipole and Raman transitions for a system with D_3 symmetry.*

D_3	Infra-red active	Raman active
Γ_1	no	yes
Γ_2	yes	no
Γ_3	yes	yes

The form of the components of $\boldsymbol{\alpha}$ can now be established using the orthogonality theorem quoted above. We have, for $|\nu_0\rangle = |\Gamma_1^{(i)} >$ and $|\nu\rangle = |\Gamma_1^{(f)} >$,

$$\begin{aligned}
\tfrac{1}{2}\,\langle\Gamma_1^{(f)}|(\alpha_{xx}+\alpha_{yy})|\Gamma_1^{(i)}\rangle &= p \neq 0\ ,\\
\tfrac{1}{2}\,\langle\Gamma_1^{(f)}|(\alpha_{xx}-\alpha_{yy})|\Gamma_1^{(i)}\rangle &= 0\ ,\\
\langle\Gamma_1^{(f)}|\alpha_{zz}|\Gamma_1^{(i)}\rangle &= q \neq 0\ ,\\
\langle\Gamma_1^{(f)}|\alpha_{xy}|\Gamma_1^{(i)}\rangle = \langle\Gamma_1^{(f)}|\alpha_{yz}|\Gamma_1^{(i)}\rangle = \langle\Gamma_1^{(f)}|\alpha_{zx}|\Gamma_1^{(i)}\rangle &= 0\ .
\end{aligned}$$

Thus, for a Raman transition between initial and final states of symmetry $\Gamma_1, \boldsymbol{\alpha}$ is of the form

$$\boldsymbol{\alpha} = \begin{pmatrix} p\,o\,o \\ o\,p\,o \\ o\,o\,q \end{pmatrix} , \quad (\Gamma_1 \to \Gamma_1) . \tag{7.25}$$

This means that scattering is only possible when the scattered radiation is analyzed in the direction of polarization of the incident radiation.

For Raman scattering to a doublet of symmetry Γ_3 with components of symmetry X or Y we obtain in a similar manner

$$\begin{aligned} \langle X|\tfrac{1}{2}\,(\alpha_{xx}+\alpha_{yy})|\Gamma_1\rangle &= 0 \ , \\ \langle X|\tfrac{1}{2}\,(\alpha_{xx}-\alpha_{yy})|\Gamma_1\rangle &= r \ , \\ \langle X|\alpha_{zz}|\Gamma_1\rangle &= 0 \ , \\ \langle X|-\alpha_{xy}|\Gamma_1\rangle &= 0 \ , \\ \langle X|\alpha_{yz}|\Gamma_1\rangle &= s \ , \qquad \text{and} \\ \langle X|-\alpha_{zx}|\Gamma_1\rangle &= 0 \ . \end{aligned}$$

Thus, for a transition $\Gamma_1 \to X$,

$$\boldsymbol{\alpha} = \begin{pmatrix} r & o & o \\ o & -r & s \\ o & s & o \end{pmatrix} , \quad (\Gamma_1 \to X) . \tag{7.26}$$

To obtain the form of $\boldsymbol{\alpha}$ for $\Gamma_1 \to Y$ we proceed in a similar fashion. However, the parameters involved are, by reason of symmetry, equal to those in Eq. (7.26) except for their positions and signs. To find them we can proceed as above. But we can also use the invariance of the matrix elements in Eq. (7.26) under the action of the symmetry operation. For example, applying the operation C_3 we find

$$\begin{aligned} r = \langle X|\tfrac{1}{2}\,(\alpha_{xx}-\alpha_{yy})|\Gamma_1\rangle &= \langle -\tfrac{1}{2}\,X - \tfrac{\sqrt{3}}{2}\,Y| - \tfrac{1}{4}\,(\alpha_{xx}-\alpha_{yy}) - \tfrac{\sqrt{3}}{2}\,(-\alpha_{xy})|\Gamma_1\rangle \\ &= \frac{r}{4} + \frac{3}{4}\,\langle Y|-\alpha_{xy}|\Gamma_1\rangle \ . \end{aligned}$$

Hence

$$\langle Y|-\alpha_{xy}|\Gamma_1\rangle = r \ .$$

We find

$$\boldsymbol{\alpha} = \begin{pmatrix} o & -r & -s \\ -r & o & o \\ -s & o & o \end{pmatrix} , \quad (\Gamma_1 \to Y) . \tag{7.27}$$

We conclude that Raman transitions from a Γ_1 ground state to a doublet Γ_3 can occur with all polarizations except that in which the incident radiation is polarized along the optic axis and the scattered radiation analyzed in the same direction.

In the presentation of experimental data the scattering geometry is denoted by a symbol of the form $\hat{\boldsymbol{k}}(\hat{e}\hat{e}')\hat{\boldsymbol{k}}'$. For example $\boldsymbol{x}(\boldsymbol{zz})\boldsymbol{y}$ means a geometry in which the incident radiation propagates along the $\boldsymbol{x}$-axis and is polarized parallel to the $\boldsymbol{z}$-axis and the scattered radiation is observed along the $\boldsymbol{y}$-axis and is analyzed parallel to the $\boldsymbol{z}$-axis. The $\boldsymbol{x}(\boldsymbol{zz}+\boldsymbol{zx})\boldsymbol{y}$ symbol described the same situation but the scattered radiation is not analyzed.

The internal motions of system could be electronic excitations as well as vibrational modes called phonons when the system of interest is a crystal.

If the primitive cell of a crystal contains two or more atoms the system possesses not only "acoustical phonons" in which the frequency approaches zero as the wavelength becomes infinitely large, but also "optical phonons," whose frequencies, at zero wave number, are different from zero; these modes correspond to the internal vibrations exhibited by a "molecule" made up by the atoms in the primitive cell.

The crystals of the diamond structure have two atoms per primitive cell and a three-fold degenerate optical phonon at the Brillouin zone center. Thus, we expect the first order Raman spectrum of diamond to exhibit a single line. In principle one could see the splitting of the line because we observe the phonon near but not exactly at $\boldsymbol{q} = 0$. However, since the optical phonon frequency near $\boldsymbol{q} = 0$ is a quadratic function of the wave vector, this separation cannot, at present, be resolved. We shall show later that in certain crystals the optical phonon frequency may vary linearly with wave vector making this dependence accessible to observation.

We consider next CaF_2 whose structure is face-centered cubic with a basis of a Ca atom at 0,0,0 and F atoms at $\frac{1}{4}, \frac{1}{4}, \frac{1}{4}$ and $-\frac{1}{4}, -\frac{1}{4}, -\frac{1}{4}$. As expected, CaF_2 has two zone center optical phonons, one of which is infra-red active while the other is Raman active.

In order to determine which phonons can be excited in a Raman process we need to establish their symmetry classifications. We limit ourselves to phonons of zero wave vector. It is therefore appropriate to give the method by means of which this classification is obtained. It is best to describe first the procedure for a molecule. Let us consider the benzene molecule (C_6H_6). The group of this molecule is D_{6h} whose character table is given in Table 7.3. The even and odd representations are designated by the subindices g and u, respectively (from the German words gerade and ungerade for even and odd). We follow the notation used in the chemical literature.[9] To determine the symmetry properties of the modes of vibration of the molecule we notice that the displacements of the 12 atoms can be described by 36 coordinates. Under an operation of the symmetry group, these are transformed so that to each operation there corresponds a matrix with 36 rows and 36 columns. As we have said before, we only require the traces of these matrices. At first sight it appears that the task of calculating these traces is a formidable one. This is, however, not the case as we now show. We first remark that if an atom does not remain fixed under a particular operation, the matrix elements corresponding to that atom are off the main diagonal and thus contribute zero to the trace. If an atom remains fixed under a transformation it contributes to the trace a number equal to the trace of the 3×3 transformation of a polar vector. This quantity is easy to find and is usually indicated in the character tables available in the literature.

Table 7.3 *Character table of* $\mathbf{D_{6h}}$ *and calculation of the symmetry of the vibrational modes of the benzene molecule.*

D_{6h}	E	C_2	$2C_3$	$2C_6$	$3C_2'$	$3C_2''$	i	iC_2	$2iC_3$	$2iC_6$	$3iC_2'$	$3iC_2''$	Basis functions
A_{1g}	1	1	1	1	1	1	1	1	1	1	1	1	
A_{1u}	1	1	1	1	1	1	−1	−1	−1	−1	−1	−1	
A_{2g}	1	1	1	1	−1	−1	1	1	1	1	−1	−1	S_z
A_{2u}	1	1	1	1	−1	−1	−1	−1	−1	−1	1	1	z
B_{1g}	1	−1	1	−1	1	−1	1	−1	1	−1	1	−1	
B_{1u}	1	−1	1	−1	1	−1	−1	1	−1	1	−1	1	
B_{2g}	1	−1	1	−1	−1	1	1	−1	1	−1	−1	1	
B_{2u}	1	−1	1	−1	−1	1	−1	1	−1	1	1	−1	
E_{1g}	2	−2	−1	1	0	0	2	−2	−1	1	0	0	S_x, S_y
E_{1u}	2	−2	−1	1	0	0	−2	2	1	−1	0	0	x, y
E_{2g}	2	2	−1	−1	0	0	2	2	−1	−1	0	0	
E_{2u}	2	2	−1	−1	0	0	−2	−2	1	1	0	0	

Fixed atoms	12	0	0	0	4	0	0	12	0	0	0	4
V	3	−1	0	2	−1	−1	−3	1	0	−2	1	1
Γ_T	36	0	0	0	−4	0	0	12	0	0	0	4
R	3	−1	0	2	−1	−1	3	−1	0	2	−1	−1
$V+R$	6	−2	0	4	−2	−2	0	0	0	0	0	0
Γ_{int}	30	2	0	−4	−2	2	0	12	0	0	0	4
$V \times V$	9	1	0	4	1	1	9	1	0	4	1	1
$[V \times V]$	6	2	0	2	2	2	6	2	0	2	2	2

Table 7.3 shows the number of fixed atoms under each transformation. The row designated by V is the representation generated by the components of a polar vector (x, y, z) *i.e.*,

$$V = A_{2u} + E_{1u} \; . \tag{7.28}$$

The row Γ_T is the 36-dimensional representation generated by the displacements of all 12 atoms in the molecule and is obtained by simply multiplying the two previous rows. R is the representation generated by a pseudovector with components S_x, S_y, S_z. We now remember that of the 36 degrees of freedom of the molecule 3 correspond to a translation of the center of mass and another 3 to rigid rotations of the molecule. Since rotations can be represented by a pseudo-vector these 6 degrees of freedom give the representation $V + R$. The representation generated by the degrees of freedom associated with the vibrational modes is then

$$\Gamma_{\text{int}} = \Gamma_T - (V + R) \tag{7.29}$$

(Γ_{int} signifies the representation of the "internal" degrees of freedom).

The reduction of Γ_{int} gives

$$\Gamma_{\text{int}} = 2A_{1g} + A_{2g} + A_{2u} + 2B_{1u} + 2B_{2g} + 2B_{2u} + E_{1g} + 3E_{1u} + 4E_{2g} + 2E_{2u} \,. \tag{7.30}$$

Of these only A_{2u} and E_{1u} are infra-red active, so that of the 20 frequencies in the vibrational spectrum only four are observed in the optical absorption spectrum. To find which among the 20 frequencies are Raman active we must find the representation of D_{6h} generated by the components of the symmetric second rank tensor $\boldsymbol{\alpha}$. This is clearly the representation obtained by removing the representation R from that generated by the functions $V_i V_j'$ $(i, j = x, y, z)$ where $\boldsymbol{V}$ and $\boldsymbol{V}'$ are polar vectors. The representation generated by the nine functions $V_i V_j'$ is called $V \times V$ and its character table is easily shown to be composed of the squares of the traces of that of V. We define

$$[V \times V] = V \times V - R \,. \tag{7.31}$$

It follows that

$$[V \times V] = 2A_{1g} + E_{1g} + E_{2g} \,, \tag{7.32}$$

so that of all the 20 frequencies only seven are observed in the Raman spectrum, namely the $2A_{1g}$, E_{1g} and the $4E_{2g}$.

The analysis for the zone center phonons in CaF_2 is shown in Table 7.4. The only difference is that now we do not subtract the rotational modes and that care must be taken in finding the number of atoms which remain fixed by considering equivalent atoms in different cells as being the same. This table also shows the analysis for an octahedral molecule of the form XY_6 in which there is an X atom at the center and a Y atom at each vertex of a regular octahedron. The details are left to the reader.[11]

We now describe the theory of some experiments showing an interesting angular dependence of the first order Raman effect in crystals. We remember that in the first order Raman effect, light of wave vector $\boldsymbol{k}$ impinges on a crystal, is scattered with wave vector $\boldsymbol{k}'$ while the crystal changes its vibrational state by having one more (or one less) phonon of wave vector $\boldsymbol{q}$. The periodicity of the crystal requires that the phonon excitation be such that the wave vector is conserved, *i.e.*,

$$\boldsymbol{k} = \boldsymbol{k}' \pm \boldsymbol{q} \,.$$

We recall that for scattering of visible radiation ($k \sim 10^5$ cm^{-1}), if the Raman shift is of the order of 100 meV or less, $|\boldsymbol{k}'| \cong |\boldsymbol{k}|$, then the transfer of momentum $\boldsymbol{q}$ has a magnitude

$$|\boldsymbol{q}| \approx 2|\boldsymbol{k}| \sin \frac{\theta}{2}$$

Table 7.4 *Character table of* O_h *and calculation of the symmetry of the vibrational modes of* CaF_2 *and of the* XY_6 *molecule.*

O_{6h}	E	$8C_3$	$3C_2 = 3C_4^2$	$6C_2$	$6C_4$	i	$8iC_3$	$3iC_2$	$6iC_2$	$6iC_4$	Basis functions
A_{1g}	1	1	1	1	1	1	1	1	1	1	
A_{2g}	1	1	1	–1	–1	1	1	1	–1	–1	
E_g	2	–1	2	0	0	2	–1	2	0	0	
T_{1g}	3	0	–1	–1	1	3	0	–1	–1	1	S_x, S_y, S_z
T_{2g}	3	0	–1	1	–1	3	0	–1	1	–1	
A_{1u}	1	1	1	1	1	–1	–1	–1	–1	–1	
A_{2u}	1	1	1	–1	–1	–1	–1	–1	1	1	
E_u	2	–1	2	0	0	–2	1	–2	0	0	
T_{1u}	3	0	–1	–1	1	–3	0	1	1	–1	x, y, z
T_{2u}	3	0	–1	1	–1	–3	0	1	–1	1	
CaF_2											
Fixed atoms	3	3	3	1	1	1	1	1	3	3	
Γ_T	9	0	–3	–1	1	–3	0	1	3	–2	
Γ_{int}	6	0	–2	0	0	0	0	0	2	–2	$= T_{1u} + T_{2g}$
XY_6											
Fixed atoms	7	1	3	1	3	1	1	5	3	1	
Γ_T	21	0	–3	–1	3	–3	0	5	3	–1	
Γ_{int}	15	0	–1	1	1	–3	0	5	3	–1	
$[V \times V]$	6	0	2	2	0	6	0	2	2	0	

where θ is the angle of scattering. Thus $|\boldsymbol{q}| \approx 10^5$ cm^{-1}, and the internal excitation of the system has a wave number close to $|\boldsymbol{q}| \approx 0$, *i.e.*, at the center of the Brillouin zone.

The energy shift of the scattered radiation with respect to the incident radiation is thus a measure of the phonon energies near the center of the Brillouin Zone. Depending on the complexity of the crystal under investigation, a few or many Raman lines can be observed. If we imagine that the effect is observed in different directions these shifts are expected to be independent of orientation. Of course, when we are dealing with Brillouin scattering where the scattering takes place with emission or absorption of an acoustic phonon this is not the case. However, even when an optical phonon is involved we expect an angular dependence. This arises because of two reasons. In an ionic crystal the polarization associated with a LO phonon gives rise to an electric field which can remove the degeneracy of the LO and TO modes. There is an angular dependence since this field may be present for propagation in some directions but not on others.

Another origin of angular dependence is the variation of the phonon frequency with wave vector. This is in general quadratic in the wave vector. However, for degenerate modes of crystal classes which do not contain the inversion operation or other improper operations, it is possible to have a linear wave vector dependence. The phenomenon can, of course, be more easily accessible in such materials where the Raman line has sufficiently small linewidth.

The phonon modes exhibiting such linear $\boldsymbol{q}$-vector dependences are:

1. the doubly degenerate phonons of crystals of symmetry

$$C_4, D_4, C_3, D_3, C_6, D_6 \; .$$

2. the triply-degenerate phonons of crystals of classes

$$T \;\; , \;\; O \; .$$

As an example we consider α-quartz.

The form of the $\omega(\boldsymbol{q})$ relations for optical phonons near $\boldsymbol{q} = 0$ is expected to satisfy

$$\omega(\boldsymbol{q}) = \omega(-\boldsymbol{q}) \tag{7.33}$$

from time reversal symmetry so that for a non-degenerate optical phonon at $\boldsymbol{q} = 0$,

$$\omega(\boldsymbol{q}) = \omega_0 + A_x q_x^2 + A_y q_y^2 + A_z q_z^2 + \dots \; . \tag{7.34}$$

In α-quartz we should have

$$\omega(\boldsymbol{q}) = \omega_0 + A\big(q_x^2 + q_y^2\big) + B q_z^2 + \dots \; . \tag{7.35}$$

We explore now in more detail the consequences of time reversal symmetry. Let us imagine a classical system starting at position $\boldsymbol{r}_i(0)$ with momenta $\boldsymbol{p}_i(0)$ and reaching at time t position $\boldsymbol{r}_i(t)$ with momenta $\boldsymbol{p}_i(t)$. If at that time all velocities are suddenly reversed and the system is allowed to proceed according to its natural motion, we say that there is time reversal symmetry if the system reaches $\boldsymbol{r}_i(0)$ with momenta $-\boldsymbol{p}_i(0)$ at time $2t$. Thus, if we start the system as above and at time t we reverse the velocities and start counting the time backwards, the system retraces the same path as before but in reverse order (see Fig. 7.2).

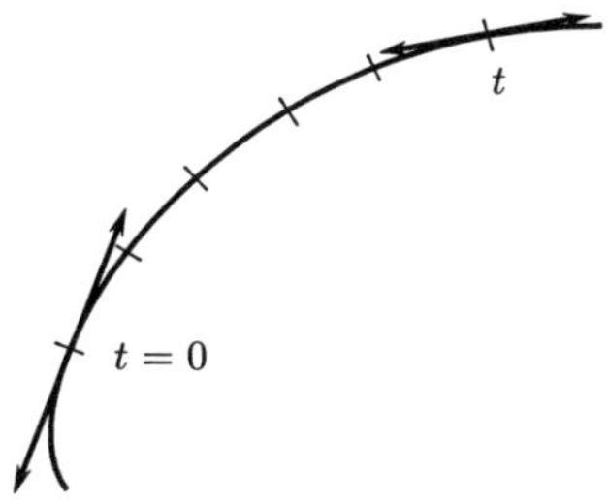

Figure 7.2 A system having time reversal symmetry.

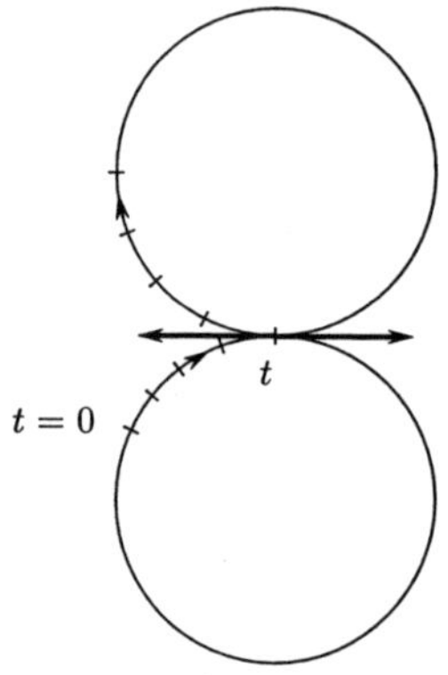

Figure 7.3 A system which does not obey time reversal invariance; charged particle in a magnetic field.

For a charged particle in a magnetic field this symmetry obviously does not hold. This is shown in Fig. 7.3.

From the mathematical point of view this says that the trajectories are the same if we change the sense of time; *i.e.*, if we make the transformation $t \to t' = -t$. A set of particles subject to a potential which depends only on the positions of the particles but not on their velocities satisfies time reversal symmetry. If $\phi(\boldsymbol{r}_1, \boldsymbol{r}_2, \ldots \boldsymbol{r}_N)$ is the potential and M_i the mass of the *ith* particle,

$$M_i \frac{d^2\boldsymbol{r}_i}{dt^2} = -\nabla_i \phi \tag{7.36}$$

is invariant under time reversal.

Let us consider now the motion of the atoms in a crystal. In each cell there are f atoms ($f = 9$ for α-quartz). The positions of their coordinates are designated by $x_{n\alpha}$ where $\boldsymbol{n}$ goes over all primitive cells, and $\alpha = 1, 2, 3, \ldots 3f$ designates the $3f$ coordinates of the f particles in the *nth* cell. The equation of motion is

$$M_\alpha \ddot{x}_{n\alpha} = -\sum_{n'\beta} C_{\alpha\beta}(\boldsymbol{n} - \boldsymbol{n}')x_{n'\beta} \; . \tag{7.37}$$

The $C_{\alpha\beta}(\boldsymbol{n} - \boldsymbol{n}')$ are real quantities which depend only on $\boldsymbol{n} - \boldsymbol{n}'$ to satisfy the translational symmetry. If we attempt to find solutions of these equations of the form

$$x_{n\alpha} = M_\alpha^{-1/2} u_\alpha \exp(i(\boldsymbol{q} \cdot \boldsymbol{n} - \omega t)) \; , \tag{7.38}$$

we obtain

$$\begin{aligned} -M_\alpha^{1/2}\omega^2 u_\alpha &= \sum_{n'\beta} C_{\alpha\beta}\Big(\boldsymbol{n} - \boldsymbol{n}'\Big) M_\beta^{-1/2} u_\beta \exp(-i\boldsymbol{q} \cdot (\boldsymbol{n} - \boldsymbol{n}')) \\ &= -\sum_\beta \overline{C}_{\alpha\beta}(\boldsymbol{q}) u_\beta M_\alpha^{1/2} \; , \end{aligned} \tag{7.39}$$

or

$$\sum_{\beta} \left[\overline{C}_{\alpha\beta}(\boldsymbol{q}) - \omega^2 \delta_{\alpha\beta}\right] u_\beta = 0 \,. \tag{7.40}$$

Here

$$\overline{C}_{\alpha\beta}(\boldsymbol{q}) = \sum_{n'} \left(M_\alpha M_\beta\right)^{-1/2} C_{\alpha\beta}\left(\boldsymbol{n} - \boldsymbol{n}'\right) \exp(-i\boldsymbol{q} \cdot (\boldsymbol{n} - \boldsymbol{n}')) \,, \tag{7.41}$$

is independent of $\boldsymbol{n}$. It can be regarded as the Fourier transform of the force constants $C_{\alpha\beta}(\boldsymbol{n} - \boldsymbol{n}')$. It is easy to convince oneself that the matrix $\overline{C}_{\alpha\beta}(\boldsymbol{q})$ is Hermitian. This follows from the relation $C_{\alpha\beta}(\boldsymbol{n} - \boldsymbol{n}') = C_{\beta\alpha}(\boldsymbol{n}' - \boldsymbol{n})$. Consider now

$$x^*_{n\alpha} = M_\alpha^{-1/2} u^*_\alpha \exp(-i(\boldsymbol{q} \cdot \boldsymbol{n} - \omega t)) \,. \tag{7.42}$$

Making the changes $t \to -t$ and $\boldsymbol{q} \to -\boldsymbol{q}$ we find that $x^*_{n\alpha}$ is transformed into

$$M_\alpha^{-1/2} u^*_\alpha \exp(i(\boldsymbol{q} \cdot \boldsymbol{n} - \omega t)) \,. \tag{7.43}$$

If we are dealing with the solution of the secular equation (7.40) in the form

$$\overline{C}(\boldsymbol{q}) u = \omega^2 u \,, \tag{7.44}$$

where we designate by u a vector with the $3f$ components u_α, u^* is the time reversed solution and satisfies

$$\overline{C}^*(-\boldsymbol{q}) u^* = \omega^2 u^* \,. \tag{7.45}$$

Thus, as in quantum mechanics, the time reversed motion is described by u^* and time reversal invariance implies that

$$\overline{C}^*(-\boldsymbol{q}) = \overline{C}(\boldsymbol{q}) \,. \tag{7.46}$$

We write

$$\overline{C}(\boldsymbol{q}) = \overline{C}_0 + q_x S_x + q_y S_y + q_z S_z + \dots \,. \tag{7.47}$$

For α-quartz, this is a 27×27 matrix but, for $\boldsymbol{q} = 0$, as we have seen, $\overline{C}_0$ can be factored into nine 1×1 matrices and nine 2×2 matrices. It is enough to consider only one of these submatrices, say, that corresponding to a Γ_3 mode. Let X and Y be the two Γ_3 modes. We will show now that S_x, S_y, S_z behave in a very reasonable way. In fact

$$S_x X = \tfrac{1}{2}\left(S_x X + S_y Y\right) + \tfrac{1}{2}\left(S_x X - S_y Y\right) \,, \tag{7.48}$$

so that

$$\langle X | S_x | X \rangle \neq 0 \,, \tag{7.49}$$

and

$$\langle Y | S_x | X \rangle = \langle X | S_x | Y \rangle = 0 \,, \tag{7.50}$$

since the first term on the right hand side of Eq. (7.48) belongs to Γ_1 while the second belongs to the X-component of Γ_3. Similarly

$$\langle Y|S_x|Y\rangle = \langle Y|\tfrac{1}{2}\,(S_xY + S_yX) + \tfrac{1}{2}\,(S_xY - S_yX)\rangle = -\langle X|S_x|X\rangle\ . \tag{7.51}$$

Therefore

$$S_x = A'\begin{bmatrix}1 & 0\\ 0 & -1\end{bmatrix} = A'\Sigma_z\ . \tag{7.52}$$

In an analogous fashion we can show that

$$S_y = -A'\begin{bmatrix}0 & 1\\ 1 & 0\end{bmatrix} = -A'\Sigma_x\ , \tag{7.53}$$

and

$$S_z = B'\begin{bmatrix}0 & -i\\ i & 0\end{bmatrix} = B'\Sigma_y\ . \tag{7.54}$$

Here A' and B' are real since S_x, S_y and S_z must be Hermitian. We have seen above that the action of the time reversal operation on the dynamical matrix $\overline{C}(\boldsymbol{q})$ is equivalent to changing $\boldsymbol{q}$ into $-\boldsymbol{q}$ and taking the complex conjugate. Thus, if Θ is the time reversal operator,

$$\begin{aligned}\Theta\overline{C}(\boldsymbol{q})\Theta^{-1} &= \Theta\overline{C}_0\Theta^{-1} + \Theta q_x\Theta^{-1}\Theta S_x\Theta^{-1} + \Theta q_y\Theta^{-1}\Theta S_y\Theta^{-1}\\ &\quad+\Theta q_z\Theta^{-1}\Theta S_z\Theta^{-1} + \dots\ .\end{aligned} \tag{7.55}$$

Using the results obtained above, we obtain

$$\Theta\overline{C}(\boldsymbol{q})\Theta^{-1} = \overline{C}^*(-\boldsymbol{q}) = \overline{C}(\boldsymbol{q}) = \overline{C}_0 - q_xS_x - q_yS_y + q_zS_z + \dots\ . \tag{7.56}$$

This result clearly implies that $A' \equiv 0$. Hence we find

$$\overline{C}(\boldsymbol{q}) = \overline{C}_0 + B'q_z\Sigma_y\ . \tag{7.57}$$

This allows us to show that $\omega(\boldsymbol{q})$ is linear near $\boldsymbol{q} = 0$. In fact, the secular equation for the 2×2 submatrix corresponding to an Γ_3 mode near $\boldsymbol{q} = 0$ is

$$\|\ \overline{C}_0 + B'q_z\Sigma_y - \omega^2\Sigma_0\ \| = 0\ . \tag{7.58}$$

where Σ_0 is the unit 2×2 matrix. If ω_0 is the value of ω at $\boldsymbol{q} = 0$, $\overline{C}_0 = \omega_0^2\Sigma_0$. The secular equation above is simply

$$\begin{bmatrix}(\omega_0^2 - \omega^2) & -iB'q_z\\ iB'q_z & (\omega_0^2 - \omega^2)\end{bmatrix} = 0\ , \tag{7.59}$$

whose solutions are

$$\omega^2 = \omega_0^2 \pm B'q_z\ . \tag{7.60}$$

For small q_z

$$\omega = \omega_0 \pm \frac{B'q_z}{2\omega_0}\ . \tag{7.61}$$

Suppose now that we apply a uniaxial stress that introduces a deformation of the crystal represented by the strain field $\boldsymbol{\epsilon}$. This quantity is defined as follows. If $\boldsymbol{u}(\boldsymbol{r})$ is the displacement of a point in the crystal originally at $\boldsymbol{r}$, the stress tensor has components

$$\epsilon_{ij} = \frac{1}{2}\left(\frac{\partial u_i}{\partial x_j} + \frac{\partial u_j}{\partial x_i}\right), \quad i, j = x, y, z\ . \tag{7.62}$$

It is clearly a second rank symmetric tensor. The change in energy, to first order in the strain, of a phonon mode due to the presence of a uniaxial strain can be expressed as

$$V = \sum_{i,j} V_{ij}\epsilon_{ij}\ , \tag{7.63}$$

where the matrices V_{ij} are the components of a symmetric second rank tensor. We have seen that a second rank tensor generates the representation

$$[\epsilon] = 2\Gamma_1 + 2\Gamma_3\ . \tag{7.64}$$

This can be seen directly by noticing that

$$\begin{aligned} V = \Sigma_{ij} V_{ij}\epsilon_{ij} &= \frac{1}{2}\,(V_{xx} + V_{yy})(\epsilon_{xx} + \epsilon_{yy}) + V_{zz}\epsilon_{zz} \\ &+ \frac{1}{2}\,[(V_{xx} - V_{yy})(\epsilon_{xx} - \epsilon_{yy}) + (-2V_{xy})(-2\epsilon_{xy})] \\ &+ 2V_{yz}\epsilon_{yz} + 2(-2V_{zx})(-\epsilon_{zx})\ . \end{aligned} \tag{7.65}$$

It is left to the reader to show that the change in energy of a phonon of symmetry A_1 is of the form

$$\Delta E_{\Gamma_1} = e(\epsilon_{xx} + \epsilon_{yy}) + f\epsilon_{zz}\ , \tag{7.66}$$

where e an f are constants. The matrix representing V for an Γ_3 phonon is of the form

$$[V] = \{a(\epsilon_{xx} + \epsilon_{yy}) + b\epsilon_{zz}\}\Sigma_0 + \{c(\epsilon_{xx} - \epsilon_{yy}) + d\epsilon_{yz}\}\Sigma_z + (2c\epsilon_{xy} + d\epsilon_{zx})\Sigma_x\ . \tag{7.67}$$

The quantities a, b, c, d are constants. They are called deformation potentials since they give the change in the energy $\hbar\omega$ of a phonon per unit strain component. If we consider now the matrix representing the branches of a Γ_3 phonon mode for finite $\boldsymbol{q}$ and keeping linear terms in $\boldsymbol{q}$ and ϵ_{ij} we can express the 2×2 Hamiltonian submatrix by

$$H = H_0 + [a(\epsilon_{xx} + \epsilon_{yy}) + b\epsilon_{zz}]\Sigma_0 + \boldsymbol{\hbar}_0 \cdot \boldsymbol{\Sigma}, \tag{7.68}$$

where

$$\begin{aligned} h_x &= 2c\epsilon_{xy} + d\epsilon_{zx}\ , \\ h_y &= Bq_z\ , \end{aligned} \tag{7.69}$$

and

$$h_z = c(\epsilon_{xx} - \epsilon_{yy}) + d\epsilon_{yz}\ .$$

The analogy with the Hamiltonian for an electron spin in the presence of a magnetic field permits us to find the eigenvalues and eigenvectors of H. They are

$$E_{1,2} = \hbar\omega_0 + [a(\epsilon_{xx} + \epsilon_{yy}) + b\epsilon_{zz}] \pm |\boldsymbol{h}| \ , \tag{7.70}$$

and

$$H\varphi_i = E_i\varphi_i \quad i = 1, 2, \tag{7.71}$$

$$\varphi_1 = X \cos\left(\frac{1}{2}\ \Omega\right) + Ye^{i\varphi} \sin\left(\frac{1}{2}\ \Omega\right) \ , \tag{7.72}$$

$$\varphi_2 = -X \sin\left(\frac{1}{2}\ \Omega\right) + Ye^{i\varphi} \cos\left(\frac{1}{2}\ \Omega\right) \ . \tag{7.73}$$

In these expressions

$$\cos\Omega = \frac{h_z}{|\boldsymbol{h}|} \ , \tag{7.74}$$

and

$$e^{i\varphi} \sin\Omega = \frac{(h_x + ih_y)}{|\boldsymbol{h}|} \ ; \tag{7.75}$$

X and Y are the degenerate eigenvectors of H_0, *i.e.*,

$$H_0X = \hbar\omega_0 X \tag{7.76}$$

$$H_0Y = \hbar\omega_0 Y \ . \tag{7.77}$$

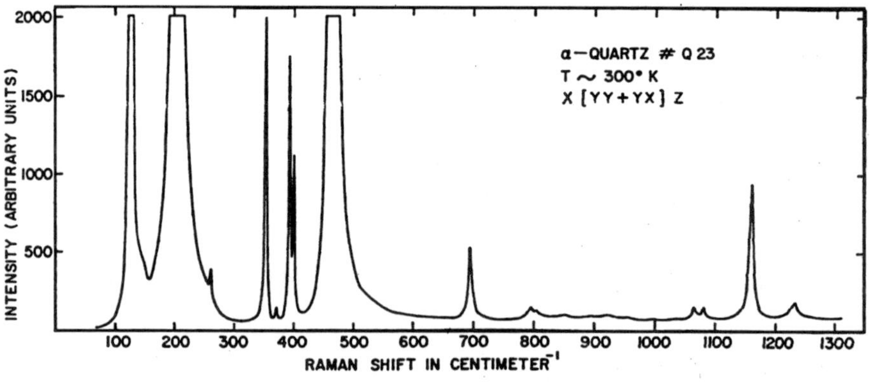

Figure 7.4 Raman spectrum of α-quartz at $T \sim 300^o$K excited by the 4880 Å Ar^+ radiation in a right angle scattering geometry. The incident light is along $\hat{x}$, the binary axis, and polarized perpendicular to the scattering plane, while the scattered radiation is along $\hat{z}$, the trigonal axis, with no analyzer in its path. The strong lines have been allowed to go out of scale to permit a clearer presentation of the weaker features. (R. J. Briggs and A. K. Ramdas, Phys. Rev. B **16**, 3815 (1997).

Figure 7.4 shows the Raman spectrum of α-quartz[10] at room temperature while Fig. 7.5 shows the same spectrum obtained at liquid helium temperature. We notice the remarkable narrowing of the lines, particularly that at about ~ 128

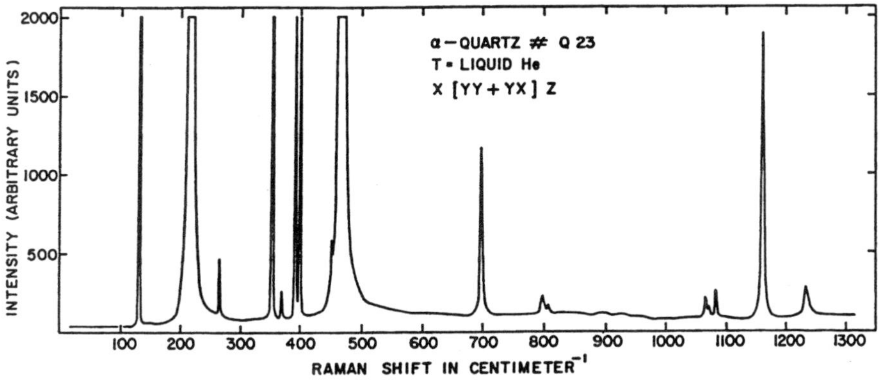

Figure 7.5 Raman spectrum of α-quartz with liquid helium as coolant. Experimental conditions same as in Fig. 7.4. *(loc. cit.* Fig. 4).

cm^{-1}. Figures 7.6 and 7.7 show the Raman spectrum of α-quartz for uniaxial strains parallel to the $\boldsymbol{x}$- and $\boldsymbol{y}$-axis, where the 128 cm^{-1} line experiences a splitting (symmetry Γ_3). To determine the splitting arising from the linear $\boldsymbol{q}$-vector dependence, measurements were performed in a crystal cut at an angle of 45° with respect to the trigonal axis. This allows us by simply reversing the direction of incidence of the laser beam to change the wave vector of the phonon involved in the process of Raman scattering. This is illustrated in Fig. 7.8. Additional measurements[12,13] are shown in Figs. 7.9-11.

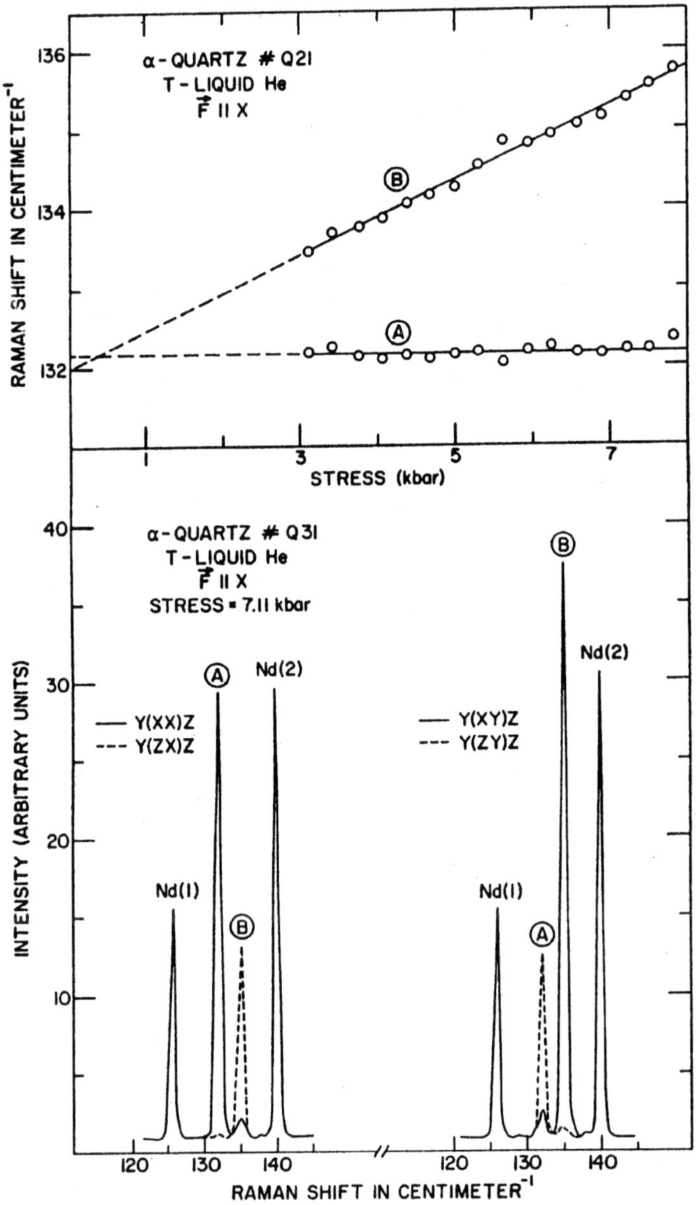

Figure 7.6 The effect of uniaxial stress on the 128 cm^{-1} Raman line of α-quartz for $\boldsymbol{F} \parallel \hat{x}$. Liquid helium used as coolant. In the upper part of the figure are shown the positions of the A and B components as functions of stress; the straight line passing through the data points for the B component represents a least squares fit while the straight line passing through the data points for the A component denotes its average position. The lower part of the figure shows the two stress induced components for the polarization parameters indicated. Stress – 7.11 kbar. Note the small residual intensity at the position of the A component for zx and xy polarizations and at the position of the B component for xx and zy polarizations can be attributed to the optical activity along $\hat{z}$. *(loc. cit.)*

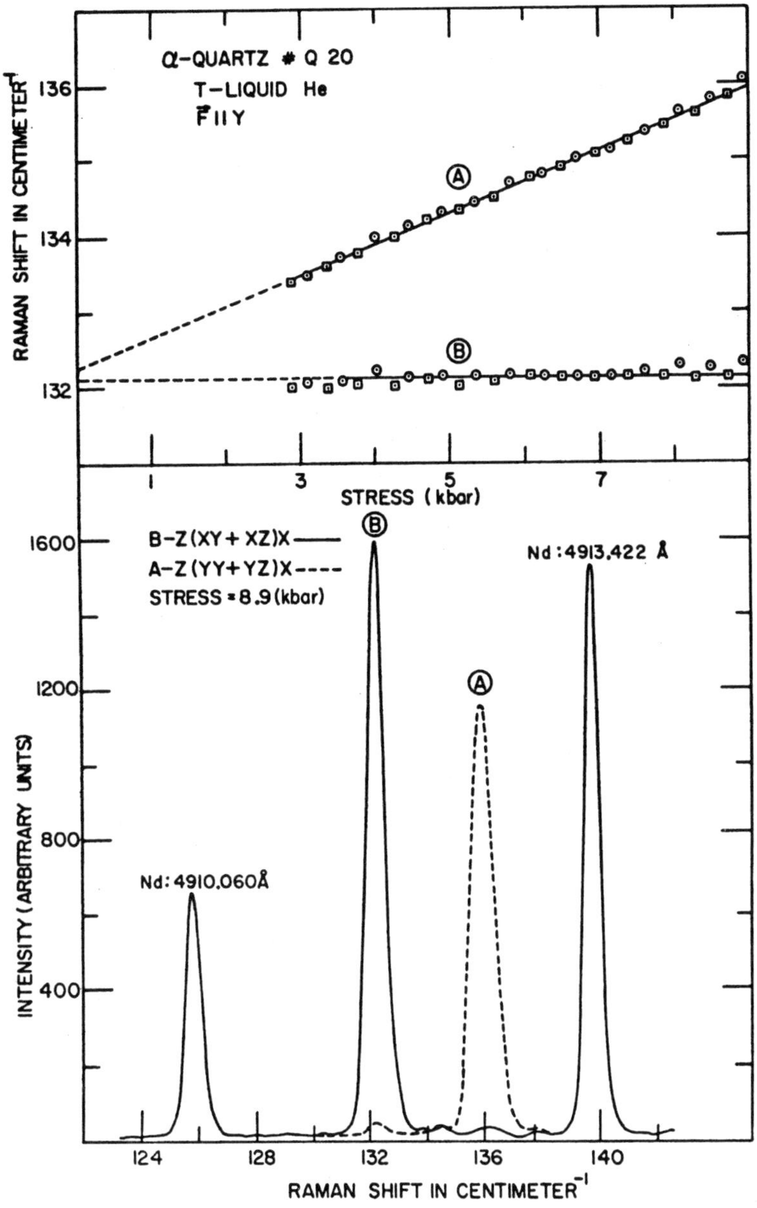

Figure 7.7 The effect of uniaxial stress on the 128 cm^{-1} Raman line of α-quartz for $\boldsymbol{F} \parallel \hat{y}$. Liquid helium used as coolant. The upper portion shows the stress dependence of the A and B components. The straight line passing through the data points for the A component represents a least squares fit while the straight line passing through the data points for the B component denotes its average position. The lower part of the figure shows the two stress induced components for the polarization parameters indicated. *loc. cit.)*

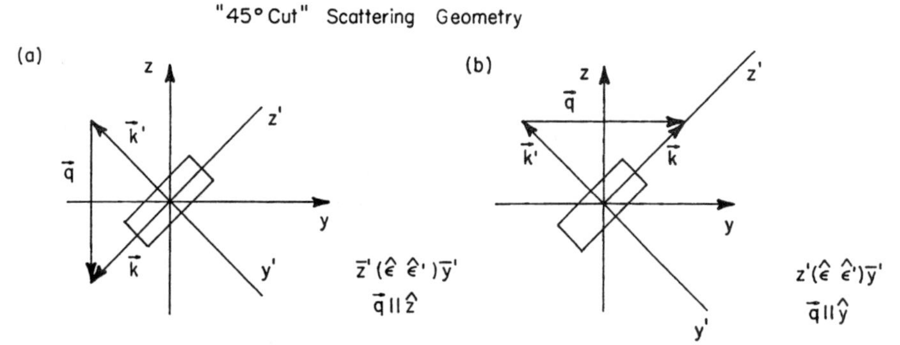

Figure 7.8 "45° cut" scattering geometry used for $\boldsymbol{F} \parallel \hat{x}$, a 2–fold axis. The force $\boldsymbol{F}$ is normal to the horizontal scattering plane and the phonon wave vector $\boldsymbol{q}$ is along $\hat{z}$, the optic axis, in (a) and along $\hat{y}$ in (b). *(loc. cit.)*

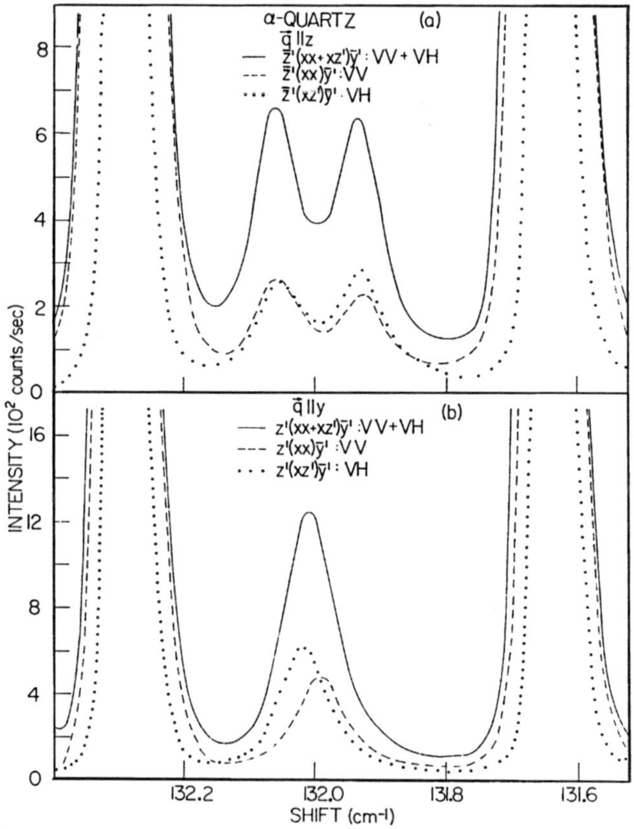

Figure 7.9 Effect of phonon propagation direction on the 128-cm^{-1} Raman line of α-quartz. (a) Scattering geometry: $\bar{z}'(xx + xz')\bar{y}'$ corresponding to $\boldsymbol{q} \parallel \hat{z}$. (b) Scattering geometry: $z'(xx+xz')\bar{y}'$ corresponding to $\boldsymbol{q} \parallel \hat{y}$. Liquid helium used as coolant. The abscissa refers to the Stokes Raman lines in the center of the figure excited with the 5145–Å Ar^+ line. The features occurring on the right and left of the Raman lines are the unshifted laser radiation appearing with an interference order difference of 203 and 204, respectively, referred to the Raman lines. (M. H. Grimsditch, A. K. Ramdas, S. Rodriguez, and V. J. Tekippe, Phys. Rev. B **15**, 5869 (1977)).)

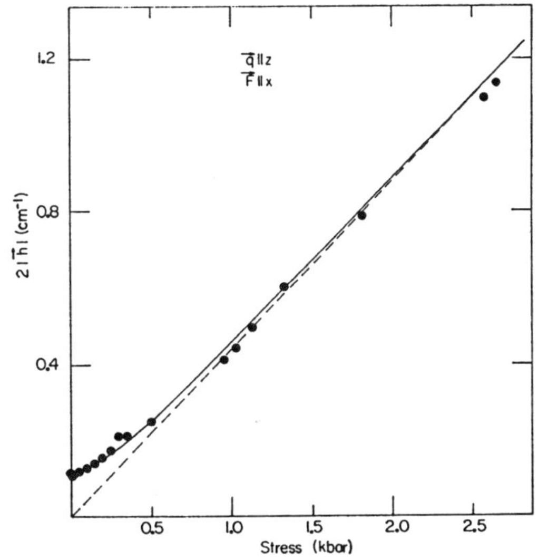

Figure 7.10 Splitting of the 128-cm^{-1} line, $2|\boldsymbol{h}|$, as a function of uniaxial stress for $\boldsymbol{F} \parallel \hat{x}$ and $\boldsymbol{q} \parallel \hat{z}$. The solid line represents a theoretical fit to the data according to theory. The dashed straight line passing through the origin has the slope obtained for $\boldsymbol{F} \parallel \hat{x}$ and $\boldsymbol{q} \parallel \hat{y}$. *(loc. cit.)*

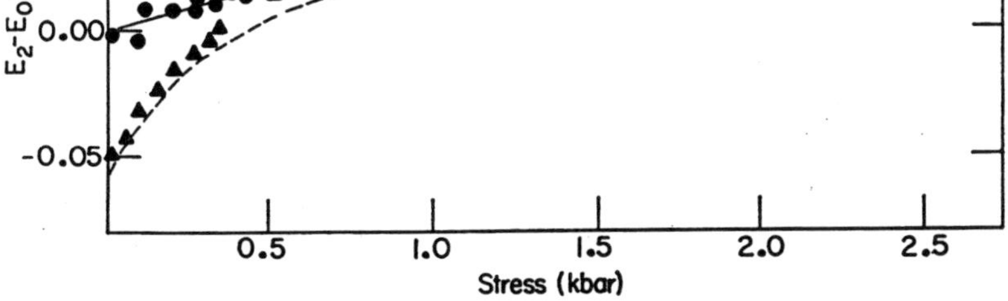

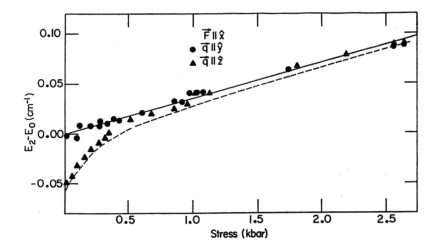

Figure 7.11 Position of the low-energy component of the 128-cm^{-1} line E_2 for $\boldsymbol{F} \parallel \hat{x}$ and $\boldsymbol{q} \parallel \hat{y}$ or $\boldsymbol{q} \parallel \hat{z}$. measured with respect to E_0, the energy of the line in the absence of stress and linear q-vector dependence. The solid line is a least-squares fit to the data for $\boldsymbol{q} \parallel \hat{y}$, whereas the dashed line represents a theoretical fit to the data for $\boldsymbol{q} \parallel \hat{z}$. *(loc. cit.)*

Chapter 8

Electronic Raman Effect: Acceptors in Group IV Elemental Semiconductors.

In Chapter 7 we have given examples of Raman scattering in which the excitations of the scattering system are vibrational modes. We now give an example of light scattering accompanied by electronic excitations. The case of boron acceptors in diamond is an example taken from our recent interests.[14–17].

The site symmetry T_d of a substitutional acceptor in the diamond structure of a group IV elemental semiconductor requires the states of a single hole bound to it to belong to one of the irreducible representations Γ_6, Γ_7, or Γ_8 of the double group $\overline{T}_d$.[10] The ground state of the hole originates from the sp^3 configuration of the acceptor as it saturates the bonds of its four nearest neighbors. The orbital ground state belongs to the Γ_5 irreducible representation of T_d. Taking the spin of the hole into account this level belongs to the $\Gamma_5 \times \Gamma_6$ representation of $\overline{T}_d$, split by the spin-orbit interaction into a quadruplet Γ_8 and a doublet Γ_7. Following von der Lage and Bethe[18] we denote states belonging to the $\Gamma_1, \Gamma_2, \Gamma_3, \Gamma_4$, and Γ_5 irreducible representations of T_d by the Greek letters α, β, γ, δ, and ϵ, respectively. Thus, we label the states of Γ_5 transforming under the action of the operations of the symmetry group as x_1, x_2, x_3 (or x, y, z), the projections of a vector on the cubic axes of the crystal, by ϵ_1, ϵ_2, ϵ_3. The angular momentum states of an atomic p-level also generate the $\Gamma_5 \times \Gamma_6$ representation of $\overline{T}_d$. This makes possible the assignment of quantum numbers to the states of the hole in one-to-one correspondence with those of the atomic states. The Γ_8 states will be denoted here by $\psi_{3/2}$, $\psi_{1/2}$, $\psi_{-1/2}$, and $\psi_{-3/2}$ where the sub-indices correspond to the rows of the representation Γ_8 generated by $p_{3/2}$ atomic states. The Γ_7 states are labeled $\phi_{1/2}$ and $\phi_{-1/2}$ in correspondence with $p_{1/2}$ atomic states. Making use of the Clebsch-Gordan coefficients for the double group $\overline{T}_d$ these states are

$$\psi_{3/2} = -(i/\sqrt{2})(\epsilon_1 + i\epsilon_2)\chi_+ \,, \tag{8.1}$$

$$\psi_{1/2} = -(i/\sqrt{6})[(\epsilon_1 + i\epsilon_2)\chi_- - 2\epsilon_3\chi_+] \,, \tag{8.2}$$

$$\psi_{-1/2} = (i/\sqrt{6})[(\epsilon_1 - i\epsilon_2)\chi_+ + 2\epsilon_3\chi_-] \,, \tag{8.3}$$

$$\psi_{-3/2} = (i/\sqrt{2})(\epsilon_1 - i\epsilon_2)\chi_- \,, \tag{8.4}$$

$$\phi_{1/2} = -(i/\sqrt{3})[(\epsilon_1 + i\epsilon_2)\chi_- + \epsilon_3\chi_+] \,, \tag{8.5}$$

and

$$\phi_{-1/2} = -(i/\sqrt{3})[(\epsilon_1 - i\epsilon_2)\chi_+ - \epsilon_3\chi_-] \,, \tag{8.6}$$

where $\chi_\pm$ are the eigenvectors of the 3-component of the spin of the hole. Even though the Hamiltonian of the bound hole lacks spherical symmetry, the quantum numbers ascribed to the states (8.1)-(8.6) have well defined physical meanings whose significance will be shown in more detail later. We call them "pseudo angular momentum quantum numbers".

The Γ_8 and Γ_7 representations of $\overline{T}_d$ are of type c in the classification of Frobenius and Schur[19] so that the subspace spanned by $\{\psi_M\}$ $(M = \pm 3/2, \pm 1/2)$ and that generated by the time-reversed states $\{\Theta\psi_M\}$ are identical. The same is true for $\{\phi_m\}$ and $\{\Theta\phi_m\}$ $(m = \pm 1/2)$. The states (8.1)-(8.6) satisfy the relations

$$\Theta\psi_{3/2} = \psi_{-3/2} \,, \tag{8.7}$$

$$\Theta\psi_{1/2} = -\psi_{-1/2} \,, \tag{8.8}$$

and

$$\Theta\phi_{1/2} = \phi_{-1/2} \,. \tag{8.9}$$

We recall that, for a single hole

$$\Theta^2 = -1 \,. \tag{8.10}$$

In the presence of a magnetic field $\boldsymbol{B}$, the new symmetry group contains the common elements of $\overline{T}_d$ and $\overline{C}_{\infty h}$, the axis of rotation of the latter being along $\boldsymbol{B}$. For $\boldsymbol{B}$ parallel to [001], [111] or [110] the symmetry group is $\overline{S}_4$, $\overline{C}_3$ or $\overline{C}_s$, respectively. For a direction of $\boldsymbol{B}$ other than a high symmetry direction of the crystal the group is $\overline{C}_1$. Character tables for $\overline{T}_d$, $\overline{S}_4$, $\overline{C}_3$, and $\overline{C}_s$ are given in Tables 8.1-8.4. The assignments of quantum numbers differ from those in Ref. 10 in that we adopt the law of transformation $\exp(-i\varphi\hat{\boldsymbol{n}}\cdot\boldsymbol{J})$ for a rotation by φ about an axis along $\hat{\boldsymbol{n}}$. $\boldsymbol{J}$ is the "angular momentum" operator.

Application of a magnetic field splits the Γ_8 and Γ_7 levels into non-degenerate states. The wave functions ψ_M and ϕ_m are approximate eigenfunctions of the Zeeman interaction for $\boldsymbol{B}$ parallel to [001]. Figure 8.1 shows the selection rules for transitions between the Zeeman levels caused by an interaction in the form of a second rank symmetric tensor as, for example, is appropriate for Raman or for electric quadrupole interactions.

Table 8.1 *Character Table of* $\overline{T}_d$.

$\bar{T}_d$	E	$\bar{E}$	$8C_3$	$8\bar{C}_3$	$3C_2$ $3\bar{C}_2$	$6\sigma_d$ $6\bar{\sigma}_d$	$6S_4$	$6\bar{S}_4$		Basis Functions
Γ_1	1	1	1	1	1	1	1	1	a	α
Γ_2	1	1	1	1	1	-1	-1	-1	a	β
Γ_3	2	2	-1	-1	2	0	0	0	a	γ_1, γ_2
Γ_4	3	3	0	0	-1	-1	1	1	a	$\delta_1, \delta_2, \delta_3$
Γ_5	3	3	0	0	-1	1	-1	-1	a	$\epsilon_1, \epsilon_2, \epsilon_3$
Γ_6	2	-2	1	-1	0	0	$\sqrt{2}$	$-\sqrt{2}$	c	χ_+, χ_-
Γ_7	2	-2	1	-1	0	0	$-\sqrt{2}$	$\sqrt{2}$	c	$\phi_{1/2}, \phi_{-1/2}$
Γ_8	4	-4	-1	1	0	0	0	0	c	$\psi_{3/2}, \psi_{1/2}, \psi_{-1/2}, \psi_{-3/2}$

Polynomials in x, y, z belonging to
(i) Γ_1 : $x^2 + y^2 + z^2$, xyz
(ii) Γ_2 : $x^4(y^2 - z^2) + y^4(z^2 - x^2) + z^4(x^2 - y^2)$
(iii) Γ_3 : $[2z^2 - x^2 - y^2; \sqrt{3}(x^2 - y^2)]$
(iv) Γ_4 : $[x(y^2 - z^2), y(z^2 - x^2), z(x^2 - y^2)]$
Also $[S_x, S_y, S_z]$ where S_x, S_y, S_z are component of a pseudovector, generate Γ_4
(v) Γ_5 : $[x, y, z]$ $[yz, zx, xy]$ $[x^3, y^3, z^3]$

Table 8.2 *Character table for the group* $\overline{S}_4$. x, y, z *refer to a system of coordinates in which* $\hat{z}$ *points along the axis of rotation of the operations* C_2 *and* S_4. *It is convenient to view* $\hat{x}$, $\hat{y}$, *and* $\hat{z}$ *as being directed along the cubic axes. The tenth column indicates the type of each representation in the classification of Frobenius and Schur.* $2\alpha_{33} - \alpha_{11} - \alpha_{22}$ *belong to* Γ_1, $\sqrt{3}(\alpha_1 - \alpha_2)$ *and* α_{12} *to* Γ_2 *while* $\alpha_{31} \mp i\alpha_{23}$ *belong to* Γ_3 *and* Γ_4, *respectively.* $\epsilon = \exp(i\pi/4) = (1+i)/\sqrt{2}$.

$\overline{S}_4$	E	$\overline{E}$	S_4^{-1}	$\overline{S_4^{-1}}$	C_2	$\overline{C}_2$	S_4	$\overline{S}_4$		Basis Functions
Γ_1	1	1	1	1	1	1	1	1	a	z^2 ; $x^2 + y^2$
Γ_2	1	1	-1	-1	1	1	-1	-1	a	z ; xy
Γ_3	1	1	i	i	-1	-1	$-i$	$-i$	b	$x + iy$
Γ_4	1	1	$-i$	$-i$	-1	-1	i	i	b	$x - iy$
Γ_5	1	-1	ϵ	$-\epsilon$	i	$-i$	ϵ^*	$-\epsilon^*$	b	χ_- ; $\psi_{3/2}$
Γ_6	1	-1	ϵ^*	$-\epsilon^*$	$-i$	i	ϵ	$-\epsilon$	b	χ_+ ; $\psi_{-3/2}$
Γ_7	1	-1	$-\epsilon$	ϵ	i	$-i$	$-\epsilon^*$	ϵ^*	b	$\psi_{-1/2}$; $\phi_{-1/2}$
Γ_8	1	-1	$-\epsilon^*$	ϵ^*	$-i$	i	$-\epsilon$	ϵ	b	$\psi_{1/2}$; $\phi_{1/2}$

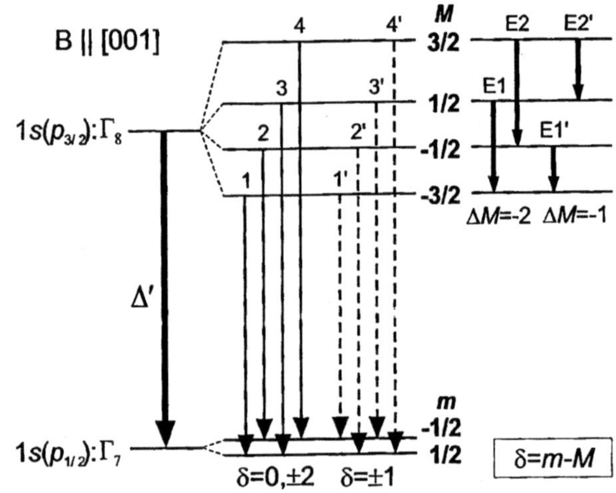

Figure 8.1 Schematic energy level diagram showing the Zeeman effect of the Δ' transition, $(1s(p_{3/2}) : \Gamma_8 \to 1s(p_{1/2}) : \Gamma_7)$ of an acceptor-bound hole in a group IV semiconductor. The figure shows the Raman allowed transitions (1, 2, 3, 4; $1', 2', 3'$ and $4'$) between the $M = 3/2, 1/2, -1/2, -3/2$ Zeeman sublevels of Γ_8 to the $m = 1/2, -1/2$ sublevels of Γ_7 corresponding to $\delta = m - M = 0, \pm 1$ and ± 2. Also shown are the Raman-EPR transitions ($E1, E2, E1'$ and $E2'$) allowed within the Γ_8 multiplet. (Ref. 17).

The order of the levels shown in Fig. 8.1 is that deduced unambiguously from an appeal to the experimental results. The left part of the figure shows transitions from initial states ψ_M to final states ϕ_m; those for which $\delta = m - M = 0, \pm 2$ are labeled 1, 2, 3, and 4 and those for $\delta = \pm 1$ are designated by the symbols $1', 2', 3', 4'$. The right hand side of the figure shows the selection rules for transitions within the Γ_8 multiplet, called here Raman electron paramagnetic resonance transitions (Raman EPR).

Those lines corresponding to a change in quantum number M by $\Delta M = -2$ are called $E1$ and $E2$ while those for which $\Delta M = -1$ are labeled $E1'$ and $E2'$. It is important to note that the transitions $3/2 \leftrightarrow -3/2$ and $1/2 \leftrightarrow -1/2$ within the Γ_8 multiplet are forbidden by time reversal symmetry. In fact, denoting by $\boldsymbol{\alpha}$ the operator causing the transition and recalling that for Raman transitions, $\boldsymbol{\alpha}$ is time-reversal invariant ($\Theta\boldsymbol{\alpha}\Theta^{-1} = \boldsymbol{\alpha}$) we have

$$< \Theta\psi|\boldsymbol{\alpha}|\psi > = < \Theta^2\psi|\boldsymbol{\alpha}|\Theta\psi >^* = - < \Theta\psi|\boldsymbol{\alpha}|\psi > \equiv 0 \qquad (8.11)$$

for any state ψ. Reference to Eqs. (8.7) and (8.8) establishes the validity of our statement. We emphasize that this restriction holds only for transitions within a single multiplet belonging to an irreducible representation of type c, not for transitions between distinct multiplets. A relaxation of this selection rule due to mixing of Γ_8 and Γ_7 states by a magnetic field has negligible experimental consequences as we shall demonstrate later.

The form of the matrix elements of $\boldsymbol{\alpha}$, the Raman tensors for the polarizability tensor $\boldsymbol{\alpha}$, can be deduced immediately from the character tables of $\overline{S_4}$, $\overline{C_3}$, and $\overline{C_s}$. For example, the tensors corresponding to transitions 1 and 2 (see Fig. 8.2),

Table 8.3 *Character table of $\overline{C}_3$. The $\hat{\zeta}$ axis is along the axis of rotation of C_3 and, for convenience in the context of this work we view $\hat{\xi}$, $\hat{\eta}$ and $\hat{\zeta}$ as being along $[11\overline{2}]$, $[\overline{1}10]$ and [111], respectively. The eighth column gives the classification of the representation according to the Frobenius and Schur test. Referring to the $\hat{\xi}$, $\hat{\eta}$, $\hat{\zeta}$ axes, $2\alpha_3 - \alpha_{11} - \alpha_2$ belongs to Γ_1, $\alpha_{31} - i\alpha_{23}$ and $\alpha_{11} - \alpha_{22} + 2i\alpha_{12}$ to Γ_2 and $\alpha_{31} + i\alpha_{23}$ and $\alpha_{11} - \alpha_{22} - 2i\alpha_{12}$ to Γ_3. $\kappa = \exp(i\pi/3)$. The basis functions ψ_M and ϕ_m have the forms given in Eqs. (1) - (6) but they refer to the $\hat{\xi}$, $\hat{\eta}$, $\hat{\zeta}$ axes.*

$\overline{C}_3$	E	$\overline{E}$	C_3	$\overline{C}_3$	C_3^{-1}	$\overline{C}_3^{-1}$		Basis Functions
Γ_1	1	1	1	1	1	1	a	ζ, S_ζ
Γ_2	1	1	κ^2	κ^2	$-\kappa$	$-\kappa$	b	$i(\xi - i\eta)$
Γ_3	1	1	$-\kappa$	$-\kappa$	κ^2	κ^2	b	$-i(\xi + i\eta)$
Γ_4	1	-1	κ	$-\kappa$	$-\kappa^2$	κ^2	b	$\psi_{-1/2}; \chi_-; \phi_{-1/2}$
Γ_5	1	-1	$-\kappa^2$	κ^2	κ	$-\kappa$	b	$\psi_{1/2}; \chi_+; \phi_{1/2}$
Γ_6	1	-1	-1	1	-1	1	a	$\psi_{3/2}; \psi_{-3/2}$

Table 8.4 *Character table of $\overline{C}_s$. The $\hat{\zeta}$ axis is normal to the reflection plane σ_h and, within the context of this work $\hat{\xi}$, $\hat{\eta}$ and $\hat{\zeta}$ are directed along $[00\overline{1}]$, $[\overline{1}10]$ and [110], respectively. The sixth column gives the Frobenius-Schur classification and, as in Table 8.3, the basis functions have the forms in (1) - (6) but are referred to the $\hat{\xi}$, $\hat{\eta}$, $\hat{\zeta}$ axes, $\alpha_{11}, \alpha_{22}, \alpha_{33}$ and α_{12} belong to Γ_1 while α_{23} and α_{31} belong to Γ_2.*

$\overline{C}_s$	E	$\overline{E}$	σ_h	$\overline{\sigma}_h$		Basis Functions
Γ_1	1	1	1	1	a	ξ, η
Γ_2	1	1	-1	-1	a	ζ
Γ_3	1	-1	i	$-i$	b	$\psi_{-3/2}; \psi_{1/2}; \chi_-; \phi_{1/2}$
Γ_4	1	-1	$-i$	i	b	$\psi_{3/2}; \psi_{-1/2}; \chi_+; \phi_{-1/2}$

when $\boldsymbol{B}$ is applied along a cubic axis (say the 3-axis) and the tensors are referred to them, are

$$\begin{pmatrix} a & ib & 0 \\ ib & -a & 0 \\ 0 & 0 & 0 \end{pmatrix}$$

and

$$\begin{pmatrix} c & 0 & 0 \\ 0 & c & 0 \\ 0 & 0 & -2c \end{pmatrix}$$

respectively. Furthermore, the quantities a and b, having the same phase, can be taken to be real without loss of generality. Clearly for line 2, c can be taken real.

Instead of using strict group-theoretical methods, additional information can be gleaned by turning to a microscopic model of the energy levels of an acceptor in an elemental group IV semiconductor. In the present context of boron in diamond, the extreme anisotropy of the valence band of diamond allows us to distinguish between symmetry allowed strong and weak transitions. To make use of this model we derive the form of the Raman scattering cross section in the limit in which the energy of the incident photon is much larger than the ionization energy of the scattering object while its wavelength is long compared to the size of the acceptor. These conditions are clearly satisfied for shallow acceptors in diamond by incident radiation in the visible region.

In this regime the differential scattering cross section for a transition from a state $|\nu_0>$ of the acceptor to a state $|\nu>$ while a photon of angular frequency ω and polarization $\hat{\mathbf{e}}$ is absorbed and replaced by one with angular frequency ω' and polarization $\hat{\mathbf{e}}'$ is given by Eq. (5.17) except that c is replaced with c/n, the velocity of light in the medium. In the high frequency limit ($\hbar\omega \gg E_i =$ ionization energy of the acceptor-bound hole) we use Eq. (5.18).

The Hamiltonian operator for the acceptor states in the field $\boldsymbol{B}$ is

$$H = H_0 + H' \tag{8.12}$$

where (in terms of hole energies)

$$H_0 = \frac{1}{m}\left(\frac{1}{2}\gamma_1 p^2 - 3\gamma_2 \sum_{i=1}^{3} p_i^2 \left(I_i^2 - \frac{1}{3}I^2\right) - 3\gamma_3 \sum_{i<j} p_i p_j \{I_i, I_j\}\right) + V(\boldsymbol{r}) \tag{8.13}$$

and

$$\begin{aligned} H' = &\frac{1}{3}\Delta' - \frac{2}{3}\Delta' \boldsymbol{I}\cdot\boldsymbol{S} + \mu_B(g_1\boldsymbol{I} + g_2\boldsymbol{S})\cdot\boldsymbol{B} + Q_1(\mu_B B)^2 \\ &+ Q_2\mu_B^2(\boldsymbol{B}\cdot\boldsymbol{I})^2 + Q_3\mu_B^2 \sum_{i=1}^{3} B_i^2 I_i^2 \ . \end{aligned} \tag{8.14}$$

The operators I_1, I_2, I_3 are angular momentum operators with angular momentum $I = 1$ in the representation generated by ϵ_1, ϵ_2 and ϵ_3, *i.e.*,

$$I_i \epsilon_j = i \sum_{k=1}^{3} \epsilon_{ijk} \epsilon_k \tag{8.15}$$

where $\epsilon_{ijk} = 1$ if ijk is an even permutation of 123, -1 if it is an odd permutation and zero in all other cases. The numbers γ_1, γ_2 and γ_3 are the Luttinger parameters,[20] $V(\boldsymbol{r})$ is the potential of the hole in the field of the ionized acceptor, $\boldsymbol{S}$ the spin operator of the hole, μ_B the Bohr magneton, g_1 and g_2, are orbital and spin g-factors, respectively; and Q_1, Q_2, Q_3 are constants (of dimensions equal to reciprocal energy). The Hamiltonian $H = H_0 + H'$ contains all possible contributions linear and quadratic in $\boldsymbol{B}$ and $\boldsymbol{p}$ consistent with the site symmetry.

Evaluation of the double commutator in Eq. (5.18) with the Hamiltonian in Eq. (8.12) gives

$$\frac{d\sigma}{d\Omega'} = \left(\frac{n^2 e^2}{mc^2}\right)^2 | < \nu | \hat{\mathbf{e}}'^* \cdot \boldsymbol{\alpha} \cdot \hat{\mathbf{e}} | \nu_0 > |^2 \tag{8.16}$$

where the tensor $\boldsymbol{\alpha}$, a dimensionless polarizability tensor, is

$$\begin{aligned}\boldsymbol{\alpha} = 3(\gamma_3 - \gamma_2)\left(2I_0^{(2)}X_0^{(2)} + \left(I_2^{(2)} + I_{-2}^{(2)}\right)\left(X_2^{(2)} + X_{-2}^{(2)}\right)\right) \\ - 6\gamma_3 \sum_{\kappa=-2}^{2} (-1)^\kappa I_\kappa^{(2)} X_{-\kappa}^{(2)} .\end{aligned} \quad (.8.17)$$

Here $I_\kappa^{(2)}$ ($\kappa = 0, \pm 1, \pm 2$) are the irreducible components of the second-rank tensor operator $\{I_i, I_j\}$, namely

$$I_0^{(2)} = \frac{1}{\sqrt{6}}(2I_3^2 - I_1^2 - I_2^2) , \quad (8.18)$$

$$I_{\pm 1}^{(2)} = \mp \frac{1}{2}\{I_3, I_1 \pm iI_2\} , \quad (8.19)$$

and

$$I_{\pm 2}^{(2)} = \frac{1}{2}(I_1^2 - I_2^2 \pm i\{I_1, I_2\}) \quad (8.20)$$

where the curly brackets stand for the anticommutator of the operators within them. The tensors $X_\kappa^{(2)}$ are the similarly defined irreducible components of the dyadics $\hat{\boldsymbol{x}}_i\hat{\boldsymbol{x}}_j + \hat{\boldsymbol{x}}_j\hat{\boldsymbol{x}}_i$ ($i, j = 1, 2, 3$) where $\hat{\boldsymbol{x}}_1, \hat{\boldsymbol{x}}_2, \hat{\boldsymbol{x}}_3$ are unit vectors along the cubic axis. For example,

$$X_{\pm 2}^{(2)} = \frac{1}{2}(\hat{\boldsymbol{x}}_1\hat{\boldsymbol{x}}_1 - \hat{\boldsymbol{x}}_2\hat{\boldsymbol{x}}_2 \pm i(\hat{\boldsymbol{x}}_1\hat{\boldsymbol{x}}_2 + \hat{\boldsymbol{x}}_2\hat{\boldsymbol{x}}_1)) . \quad (8.21)$$

The last term in Eq. (8.17), proportional to γ_3 alone, is the product of $6\gamma_3$ and the scalar product of the tensors $I^{(2)}$ and $X^{(2)}$ and is, thus, spherically symmetric, a result which will prove advantageous in the calculation of the Raman tensors for several directions of $\boldsymbol{B}$. The first term, proportional to $\gamma_3 - \gamma_2$ has only cubic symmetry, so that the departure from isotropy is measured by the difference of the Luttinger parameters γ_2 and γ_3. Equations (8.16) and (8.17) establish that the only prominent Raman transitions in the high frequency limit are those between levels belonging to the $\Gamma_5 \times \Gamma_6 = \Gamma_8 + \Gamma_7$ ground states.

We define a total "pseudo angular momentum" operator

$$\boldsymbol{J} = \boldsymbol{I} + \boldsymbol{S} \quad (8.22)$$

which corresponds to levels with $J = 3/2(\Gamma_8)$ and $J = 1/2(\Gamma_7)$. The interaction H', to first order in $\boldsymbol{B}$ becomes

$$H' \approx \frac{5}{4}\Delta' - \frac{1}{3}\Delta' J(J+1) + g_J \mu_B \boldsymbol{B} \cdot \boldsymbol{J} . \quad (8.23)$$

Application of the Wigner-Eckart theorem shows that

$$g_{3/2} = \frac{1}{3}(2g_1 + g_2) \quad (8.24)$$

and

$$g_{1/2} = \frac{1}{3}(4g_1 - g_2) \ . \tag{8.25}$$

The functions (8.1)-(8.6) are, however, not eigenvectors of H'. For $\boldsymbol{B} \parallel [001]$ the states $\psi_{1/2}$ and $\phi_{1/2}$ (belonging to Γ_8 of $\overline{S}_4$) and $\psi_{-1/2}$ and $\phi_{-1/2}$ (belonging to Γ_7 of $\overline{S}_4$) are mixed by H'. The matrix elements of H' are

$$< \psi_{\pm 3/2}|H'|\psi_{\pm 3/2} >= \pm\frac{1}{2}\mu_B B(2g_1 + g_2) + (\mu_B B)^2 Q \ , \tag{8.26}$$

$$< \psi_{\pm 1/2}|H'|\psi_{\pm 1/2} >= \pm\frac{1}{6}\mu_B B(2g_1 + g_2) + \frac{1}{3}(\mu_B B)^2 Q \ , \tag{8.27}$$

$$< \phi_{\pm 1/2}|H'|\phi_{\pm 1/2} >= \Delta' \pm \frac{\mu_B B}{6}(4g_1 - g_2) + \frac{2}{3}(\mu_B B)^2 Q \ , \tag{8.28}$$

and

$$< \psi_{\pm 1/2}|H'|\phi_{\pm 1/2} >= \sqrt{2}\, G_\pm \equiv \frac{\mu_B B\sqrt{2}}{3}(g_1 - g_2 \pm \mu_B B Q) \ . \tag{8.29}$$

Here, $Q = Q_2 + Q_3$ while the term proportional to Q_1, being the same for all six states, is omitted as it does not influence the transition energies.

The diagonalization of $H_0 + H'$ in the subspace of the functions (8.1)-(8.6) is straightforward and will not be reproduced here. The energy eigenvalues are given analytically and the mixing of states belonging to the same irreducible representation of $\overline{S}_4$ is described by

$$\phi'_{\pm 1/2} = \psi_{\pm 1/2} \sin\theta_\pm + \phi_{\pm 1/2} \cos\theta_\pm \tag{8.30}$$

and

$$\psi'_{\pm 1/2} = \psi_{\pm 1/2} \cos\theta_\pm - \phi_{\pm 12/} \sin\theta_\pm \tag{8.31}$$

where

$$\tan\theta_\pm = 2\sqrt{2}\, G_\pm \left[\Delta' \pm G_\pm + \sqrt{(\Delta' \pm G_\pm)^2 + 8G_\pm^2}\right]^{-1} \ . \tag{8.32}$$

It can be shown that the *difference in the energies of lines* $3'$ *and* $2'$ *is* $-2\mu_B B g_1$ *while the corresponding difference between lines* $4'$ *and* 1 *is* $-\mu_B B(2g_1 + g_2)$. These results were employed in the determination of the g-factors from the experimental data.

We are now in a position to calculate the Raman tensors $< \nu|\hat{\boldsymbol{e}}'^* \cdot \boldsymbol{\alpha} \cdot \hat{\boldsymbol{e}}|\nu_0 >$ for the different transitions. The mixing of states described by Eqs. (8.30) and (8.31) is taken into account by noting that

$$< \psi_{1/2}|\boldsymbol{\alpha}|\psi_{3/2} >= -\sqrt{2} < \phi_{1/2}|\boldsymbol{\alpha}|\psi_{3/2} > \tag{8.33}$$

and

$$< \psi_{1/2}|\boldsymbol{\alpha}|\psi_{-3/2} >= \frac{1}{\sqrt{2}} < \phi_{1/2}|\boldsymbol{\alpha}|\psi_{-3/2} > \tag{8.34}$$

Table 8.5 *Matrix elements of the components of the tensor operator $I^{(2)}$ between Γ_8 and Γ_7 states.*

		$\psi_{3/2}$	$\psi_{1/2}$	$\psi_{-1/2}$	$\psi_{-3/2}$
$2I_3^2 - I_1^2 - I_2^2$	$\phi_{1/2}$	0	$\sqrt{2}$	0	0
	$\phi_{-1/2}$	0	0	$-\sqrt{2}$	0
$\sqrt{3}(I_1^2 - I_2^2)$	$\phi_{1/2}$	0	0	0	$\sqrt{2}$
	$\phi_{-1/2}$	$-\sqrt{2}$	0	0	0
$\{I_2, I_3\}$	$\phi_{1/2}$	$-i/\sqrt{6}$	0	$-i/\sqrt{2}$	0
	$\phi_{-1/2}$	0	$i/\sqrt{2}$	0	$i/\sqrt{6}$
$\{I_3, I_1\}$	$\phi_{1/2}$	$-1/\sqrt{6}$	0	$1/\sqrt{2}$	0
	$\phi_{-1/2}$	0	$1/\sqrt{2}$	0	$-1/\sqrt{6}$
$\{I_1, I_2\}$	$\phi_{1/2}$	0	0	0	$-i\sqrt{(2/3)}$
	$\phi_{-1/2}$	$-i\sqrt{(2/3)}$	0	0	0

together with two other relations deduced from Eqs. (8.33) and (8.34) with the use of the time reversal operation. The matrix elements of the components of the tensor operator $I^{(2)}$ are given in Table 8.5 in a convenient way. These are the only ones that we shall need even for different directions of the magnetic field.

The Raman tensors for the eight transitions 1, 2, 3, 4, $1', 2', 3', 4'$ are given in Table 8.6 where the factors

$$f_\pm = \cos\theta_\pm \pm \frac{1}{\sqrt{2}} \sin\theta_\pm \ , \tag{8.35}$$

$$g_\pm = \cos^2\theta_\pm - \sin^2\theta_\pm \mp \frac{1}{\sqrt{2}} \sin\theta_\pm \cos\theta_\pm \ , \tag{8.36}$$

$$h_\pm = \cos\theta_\pm \mp \sqrt{2}\sin\theta_\pm \ , \tag{8.37}$$

and

$$k = \cos(\theta_+ + \theta_-) \tag{8.38}$$

reflect the effect of mixing.

In order to evaluate the Raman tensors when $\boldsymbol{B} \parallel [111]$ we proceed by transforming the wave functions by rotating the axes of coordinates to coincide with the three orthogonal directions $[11\bar{2}]$, $[\bar{1}10]$ and $[111]$ using the rotation operator $D^{(J)}(\alpha, \beta, \gamma)$ for $J = 3/2$ for the states ψ_M and $J = 1/2$ for ϕ_m. The Euler angles for this rotation are $\alpha = \pi/4$, $\beta = \arccos(1/\sqrt{3})$, $\gamma = 0$. The new wave functions can then be used to calculate the Raman tensors. A second approach, the one we follow here, consists in recognizing that the rotated functions have the same form as those in Eqs. (8.1)-(8.6) but that now $\epsilon_1, \epsilon_2, \epsilon_3$ and $\chi_\pm$ have different meanings. Instead, they now refer to the new axes. The tensor $\boldsymbol{\alpha}$, however, does not have spherical symmetry so that, in order to calculate the Raman tensors we must express it in terms of the components of the operator $\boldsymbol{I}$ on the new axes.

It is at this point that our separation of $\boldsymbol{\alpha}$ into spherically symmetric and cubic components becomes useful. The contributions to the tensors originating

Table 8.6 *The Raman tensors for transitions between the Zeeman levels of the Γ_8 manifoldto those of the spin-orbit split Γ_7 states. The magnetic field $\boldsymbol{B}$ is along [001] and the components of the tensors are referred to the cubic axes. The tensors appear in pairs of simply related tensors. In each case the upper and lower signs correspond to the upper and lower lines displayed on the left column.*

Transition	$\psi_M \to \phi_m$	Raman Tensor
1	$-3/2 \to 1/2$	$\sqrt{6}f_\pm \begin{pmatrix} \mp\gamma_2 & i\gamma_3 & 0 \\ i\gamma_3 & \pm\gamma_2 & 0 \\ 0 & 0 & 0 \end{pmatrix}$
4	$3/2 \to -1/2$	
2	$-1/2 \to -1/2$	$\pm\sqrt{2}\gamma_2 g_\mp \begin{pmatrix} -1 & 0 & 0 \\ 0 & -1 & 0 \\ 0 & 0 & 2 \end{pmatrix}$
3	$1/2 \to 1/2$	
$1'$	$-3/2 \to -1/2$	$\gamma_3 h_\mp \sqrt{(3/2)} \begin{pmatrix} 0 & 0 & 1 \\ 0 & 0 & \mp i \\ 1 & \mp i & 0 \end{pmatrix}$
$4'$	$3/2 \to 1/2$	
$2'$	$-1/2 \to 1/2$	$-(3/\sqrt{2})k\gamma_3 \begin{pmatrix} 0 & 0 & 1 \\ 0 & 0 & \mp i \\ 1 & \mp i & 0 \end{pmatrix}$
$3'$	$1/2 \to -1/2$	

from the spherical part are the same in all directions and are identical to the tensors in Table 8.6 setting $\gamma_2 = \gamma_3$. To obtain the cubic components we denote by I'_1, I'_2 and I'_3 the components of $\boldsymbol{I}$ on the new axes. Then we obtain

$$2I_3^2 - I_1^2 - I_2^2 = I_1'^2 - I_2'^2 - \sqrt{2}\{I'_3, I'_1\} \tag{8.39}$$

and

$$\sqrt{3}(I_1^2 - I_2^2) = -\sqrt{2}\{I'_2, I'_3\} - \{I'_1, I'_2\}\ . \tag{8.40}$$

The transformations of the $\hat{\boldsymbol{x}}_i\hat{\boldsymbol{x}}_j + \hat{\boldsymbol{x}}_j\hat{\boldsymbol{x}}_i$ tensors follows in the same way and have the same form as the transformations in Eqs. (8.39) and (8.40). Table 8.7 gives the tensors for the Zeeman components of the $\Gamma_8 \to \Gamma_7$ transitions referred to axes along $[11\bar{2}]$, $[\bar{1}10]$ and [111]. Two points need to be kept in mind while using the results in this table. First we note that the quadratic terms in B in H' are not the same as for $\boldsymbol{B} \parallel [001]$. In fact, the last three terms in Eq. (8.14) are

$$\left(Q_1 + \frac{2}{3}Q_3 + Q_2 I_3'^2\right)(\mu_B B)^2$$

for $\boldsymbol{B} \parallel [111]$ while they are

$$\left(Q_1 + (Q_2 + Q_3)I_3^2\right)(\mu_B B)^2$$

Table 8.7 *The Raman tensors for the Zeeman transitions of Δ' with $\boldsymbol{B} \parallel [111]$. The tensor components are referred to axes along $[11\bar{2}]$, $[\bar{1}10]$, and $[111]$. The functions $|M>$ and $|m>$ are those obtained from ψ_M and ϕ_m after rotation by the Euler angles $\alpha = \pi/4$, $\beta = \arccos 1/\sqrt{3}$, $\gamma = 0$.*

Line	$M \to m$	Raman Tensors
1	$-3/2 \to 1/2$	$f_{\pm}\sqrt{(2/3)} \begin{pmatrix} \mp(2\gamma_3+\gamma_2) & i(2\gamma_3+\gamma_2) & \mp\sqrt{2}(\gamma_3-\gamma_2) \\ i(2\gamma_3+\gamma_2) & \pm(2\gamma_3+\gamma_2) & -i\sqrt{2}(\gamma_3-\gamma_2) \\ \mp\sqrt{2}(\gamma_3-\gamma_2) & -i\sqrt{2}(\gamma_3-\gamma_2) & 0 \end{pmatrix}$
4	$3/2 \to -1/2$	
2	$-1/2 \to -1/2$	$\pm\sqrt{2}\gamma_3 g_{\mp} \begin{pmatrix} -1 & 0 & 0 \\ 0 & -1 & 0 \\ 0 & 0 & 2 \end{pmatrix}$
3	$1/2 \to 1/2$	
1′	$-3/2 \to -1/2$	$h_{\mp}/\sqrt{3} \begin{pmatrix} \gamma_3-\gamma_2 & \pm i(\gamma_3-\gamma_2) & (\gamma_3+2\gamma_2)/\sqrt{2} \\ \pm i(\gamma_3-\gamma_2) & -(\gamma_3-\gamma_2) & \mp i(\gamma_3+2\gamma_2)/\sqrt{2} \\ (\gamma_3+2\gamma_2)/\sqrt{2} & \mp i(\gamma_3+2\gamma_2)/\sqrt{2} & 0 \end{pmatrix}$
4′	$3/2 \to 1/2$	
2′	$-1/2 \to 1/2$	$k \begin{pmatrix} -(\gamma_3-\gamma_2) & \mp i(\gamma_3-\gamma_2) & -2(\gamma_3+2\gamma_2)/\sqrt{2} \\ \mp i(\gamma_3-\gamma_2) & \gamma_3-\gamma_2 & \pm i(\gamma_3+2\gamma_2)/\sqrt{2} \\ -(\gamma_3+2\gamma_2)/\sqrt{2} & \mp i(\gamma_3+2\gamma_2)/\sqrt{2} & 0 \end{pmatrix}$
3′	$1/2 \to -1/2$	

for $\boldsymbol{B} \parallel [001]$. In our development of formulas for energies, mixing, and Raman tensors we used $Q = Q_2 + Q_3$. The corresponding equations for $\boldsymbol{B} \parallel [111]$ are formally the same as for $\boldsymbol{B} \parallel [001]$ except that Q is to be replaced by Q_2. Note also that in this case the term $(Q_1 + 2Q_3/3)(\mu_B B)^2$, being the same for all six states in the $\Gamma_5 \times \Gamma_6$ manifold can be omitted since it does not influence the transition energies. Q_2 and Q_3, being small, give rise to small changes in the energy eigenvalues and the mixing parameters $\tan\theta_{\pm}$.

When $\boldsymbol{B} \parallel [110]$ the states $\psi_{1/2}$, $\phi_{1/2}$, and $\psi_{-3/2}$ belonging to Γ_3 of $\overline{C}_s$ are mixed. The same occurs for $\psi_{-1/2}$, $\phi_{-1/2}$ and $\psi_{3/2}$ belonging to Γ_4.

In order to interpret the experimental results in Ref. 17, we adopt the level ordering of the Zeeman sublevels of Γ_8 and Γ_7 hole states as displayed in Fig. 8.1. The rotational invariance of the Zeeman interaction $\mu_B(g_1\boldsymbol{I} + g_2\boldsymbol{S}) \cdot \boldsymbol{B}$ shows that the level ordering of the states characterized by $M = \frac{3}{2}, \frac{1}{2}, -\frac{1}{2}, -\frac{3}{2}$ and $m = \frac{1}{2}, -\frac{1}{2}$ is independent of the direction of $\boldsymbol{B}$ with respect to the cubic axes $< 100 >$. The quadratic terms in $\boldsymbol{B}$ in Eq. (8.14) are small and modify only slightly the separation of the sublevels. With the sign convention in Eq. (8.14), it is expected that g_2 is approximately equal to -2, the intrinsic g-factor for a free hole; similarly, g_1, the orbital g-factor, is also expected to be negative. While we temporarily adopt this view as a working hypothesis reflected in the level ordering in Fig. 8.1, we demonstrate that the experimental results for $\boldsymbol{B} \parallel [111]$

using circularly polarized light unambiguously establish the correctness of this choice.

In Fig. 8.2 we exhibit the Raman spectrum for the Zeeman splitting of Δ' with $\boldsymbol{B} \parallel [111]$ in the backscattering geometry, the incident light propagating along $\boldsymbol{B}$. The incident light is circularly polarized $\hat{\boldsymbol{\sigma}}_\pm$ and the scattered analyzed for $\hat{\boldsymbol{\sigma}}'_\pm$. Denoting by $\hat{\boldsymbol{\xi}}, \hat{\boldsymbol{\eta}}$, and $\hat{\boldsymbol{\zeta}}$ unit vectors parallel to $[11\bar{2}]$, $[\bar{1}10]$, and $[111]$, respectively, we have

$$\hat{\boldsymbol{\sigma}}_\pm = (\hat{\boldsymbol{\eta}} \pm i\hat{\boldsymbol{\xi}})/\sqrt{2} \tag{8.41}$$

and

$$\hat{\boldsymbol{\sigma}}'_\pm = (\hat{\boldsymbol{\xi}} \pm i\hat{\boldsymbol{\eta}})/\sqrt{2}\,. \tag{8.42}$$

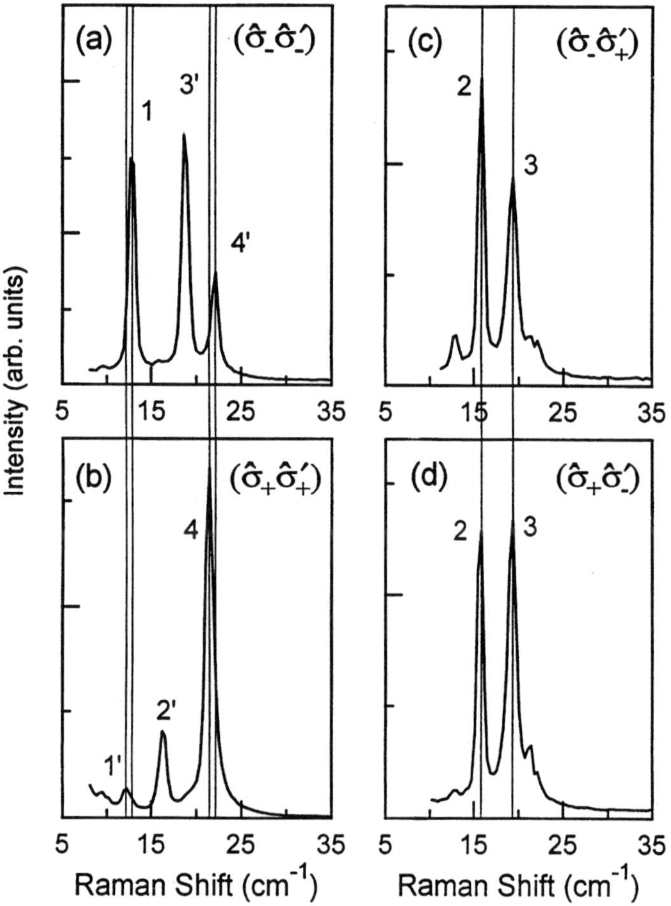

Figure 8.2 The Zeeman components of Δ', observed in backscattering along $\boldsymbol{B} \parallel [111]$, at 7 T and 5 K, the incident (scattered) light polarized (analyzed) using circular polarizations defined in Eqs. (8.41) and (8.42). The spectra excited with the 6471 Å Kr^+ laser line and recorded on a CCD-based triple grating spectrometer. The lines are labeled according to Fig. 8.1. (Ref. 17).

Table 8.8 *Theoretical Raman amplitudes of the Zeeman components of Δ' with $\mathbf{B} \parallel [111]$, observed in the backscattering geometry along $\mathbf{B}$ employing circularly polarized incident $\hat{\sigma}_\pm$ and scattered $\hat{\sigma}'_\pm$ radiation defined in the text. Amplitudes identified in boxes satisfy the conservation of 'angular momentum', i.e., sum of the change in the photon angular momentum ΔS and $\delta = m - M$. The allowed primed transitions, which do not conform to the conservation of 'angular momentum', arise from the departure from spherical symmetry of the valence band ($\gamma_2 \neq \gamma_3$).*

Line	$\Delta S \rightarrow$ / $\delta \downarrow$	$(\hat{\sigma}_+, \hat{\sigma}'_-)$ 0	$(\hat{\sigma}_+, \hat{\sigma}'_+)$ 2	$(\hat{\sigma}_-, \hat{\sigma}'_+)$ 0	$(\hat{\sigma}_-, \hat{\sigma}'_-)$ −2
1	2	0	0	0	$\boxed{2if_+(2\gamma_3 + \gamma_2)\sqrt{\frac{2}{3}}}$
2	0	$\boxed{-i\sqrt{2}\gamma_3 g_-}$	0	$\boxed{i\sqrt{2}\gamma_3 g_-}$	0
3	0	$\boxed{i\sqrt{2}\gamma_3 g_+}$	0	$\boxed{-i\sqrt{2}\gamma_3 g_+}$	0
4	−2	0	$\boxed{2if_-(2\gamma_3 + \gamma_2)\sqrt{\frac{2}{3}}}$	0	0
1′	1	0	$(2ih_-/\sqrt{3})(\gamma_3 - \gamma_2)$	0	0
2′	1	0	$-2ik(\gamma_3 - \gamma_2)$	0	0
3′	−1	0	0	0	$2ik(\gamma_3 - \gamma_2)$
4′	−1	0	0	0	$-\frac{2ih_+}{\sqrt{3}}(\gamma_3 - \gamma_2)$

The experimental results for the four configurations $\bar{\zeta}(\hat{\boldsymbol{\sigma}}_\pm, \hat{\boldsymbol{\sigma}}'_\pm)\zeta$ are displayed in Figs. 8.2(a)-8.2(d). In Fig. 8.2(a), $(\hat{\boldsymbol{\sigma}}_-, \hat{\boldsymbol{\sigma}}'_-)$, transitions 1, 3′ and 4′ appear whereas in Fig. 8.2(b), one sees 1′, 2′ and 4. Using the Raman tensors in Table 8.7, the Raman amplitudes appropriate for the four polarization configurations can be constructed as given in Table 8.8. The change in the angular momentum of the radiation upon scattering, ΔS, is 0 for $(\hat{\boldsymbol{\sigma}}_+, \hat{\boldsymbol{\sigma}}'_-)$ and $(\hat{\boldsymbol{\sigma}}_-, \hat{\boldsymbol{\sigma}}'_+)$, 2 for $(\hat{\boldsymbol{\sigma}}_+, \hat{\boldsymbol{\sigma}}'_+)$ and −2 for $(\hat{\boldsymbol{\sigma}}_-, \hat{\boldsymbol{\sigma}}'_-)$. The second column in Table 8,4 lists the change in the pseudo-angular momentum of the hole, δ, for each of the eight Zeeman components. An inspection of the table reveals that for lines 1-4, $\Delta S + \delta = 0$ consistent with conservation of "angular momentum". In contrast, lines 1′ and 2′ in

$(\hat{\sigma}_+, \hat{\sigma}'_+)$ and 3′ and 4′ in $(\hat{\sigma}_-, \hat{\sigma}'_-)$ *appear as a consequence of the departure from spherical symmetry (associated with* $\gamma_2 \neq \gamma_3$*)*. The transitions 2 and 3 allowed in both $(\hat{\sigma}_+, \hat{\sigma}'_-)$ and $(\hat{\sigma}_-, \hat{\sigma}'_+)$ conforming to $\Delta S = \delta = 0$ are shown in Figs. 8.2(c) and 8.2(d). Had we employed linearly polarized light and analyzed the scattered light linearly, the primed lines and their unprimed counterparts would appear together in the same scattering configuration preventing their clear resolution. The appearance of the eight Zeeman transitions in the four polarization configurations conform to the choice of the level ordering in Fig. 8.1. We note that the interval between 4′ and 4 is equal to that between 1 and 1′ as can be seen from the separation between the vertical lines in the figure. And since the initial states of 4′ and 4 and of 1 and 1′ are $+\frac{3}{2} : \Gamma_8$ and $-\frac{3}{2} : \Gamma_8$, respectively, clearly their spacings represent the splitting of the Γ_7 Zeeman levels. Since 4′ occurs at a higher energy than 4, the ordering of the Γ_7 Zeeman levels is that adopted in Fig. 8.1. However, had the ordering of the Γ_8 sublevels been reversed, the lines in Fig. 8.2(a) would be relabeled 4′, 3′, and 1 in order of increasing energy while in Fig. 8.2(b) they would be 4, 2′, and 1′, demonstrating that the level ordering of $-\frac{1}{2} : \Gamma_7$ and $+\frac{1}{2} : \Gamma_7$ remains the *same* as dictated by the experiment. Hence $g_{1/2} = 0.21$ is *positive.* We note that the labeling of 3′ in Fig. 8.2(a) and that of 2′ in Fig. 8.2(b) is fixed by the selection rules in Table 8.4, irrespective of the level ordering of Γ_8. The occurrence of 3′ at a higher energy compared to that of 2′ then demonstrates that the *level ordering in Fig. 8.1 is the only one consistent with experiment.* It is also noteworthy that the separation between the pairs (3, 3′) and (2′,2) *must* be equal and correspond to the Γ_7 splitting; thus a relabeling of lines 2 and 3 in Figs. 8.2(c) and 8.2(d) as 3 and 2 would result in an unacceptable contradiction.

The Δ' line at $B = 0$ exhibits a spontaneous splitting[16] of (0.8 ± 0.04) cm^{-1} attributed to a static Jahn-Teller effect. The extrapolation of the Zeeman lines to a single zero-field position at Δ' does not reflect this separation since the minimum magnetic field at which the Zeeman levels could be fully resolved is ~ 1 T. As shown in Ref. 16, the Jahn-Teller splitting results from a tetragonal distortion along $< 100 >$ which implies that at zero field the states $\pm\frac{3}{2} : \Gamma_8$ and $\pm\frac{1}{2} : \Gamma_8$ have different energies. This feature is brought out distinctly in the inset in Fig. 8.3 where the positions of the Raman-EPR lines E1, E2, E1′, and E2′ are shown. It can be shown that as B approaches zero, E2 and E2′ converge as do E1 and E1′ to different zero field positions.

The magnetic field dependence of the differences in the energies of lines 3′ and 2′ on the one hand, and of lines 4′ and 1 on the other, are linear in B as already mentioned. This allows us to obtain g_1 and g_2 from the experimental results (see Fig. 8.4).

The Raman EPR transition $\frac{1}{2}(\Gamma_8) \rightarrow -\frac{1}{2}(\Gamma_8)$ becomes weakly allowed as the applied magnetic field increases due to mixing of $\phi_{\pm 1/2}$ into $\psi_{\pm 1/2}$. For example, for $\boldsymbol{B} \parallel [001]$, the Raman tensor for this transition is

$$< \psi'_{-1/2}|\alpha|\psi'_{1/2} >= (3/\sqrt{2})\gamma_3 \sin(\theta_+ + \theta_-) \begin{pmatrix} 0 & 0 & 1 \\ 0 & 0 & i \\ 1 & i & 0 \end{pmatrix} .$$

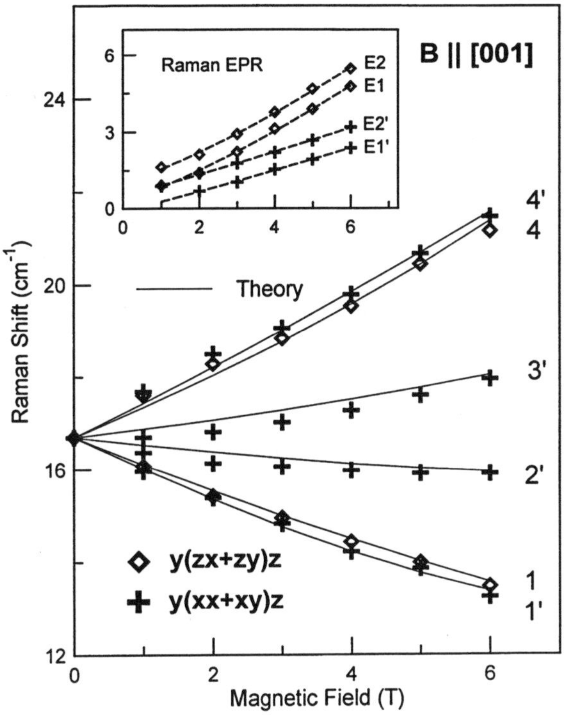

Figure 8.3 The energies of lines $1, 1', 2', 3', 4$ and $4'$ as functions of magnetic field, with $\boldsymbol{B} \parallel$ [001]; lines 2 and 3, though allowed are too weak to be observed (see text). The solid lines are theoretical, calculated using $g_1 = -0.373, g_2 = -2.112, Q_2 + Q_3 = -0.154$ cm, energy being expressed in cm^{-1}. The inset displays the magnetic field dependence of the four Raman-EPR lines. The curves in the inset passing through the experimental points, drawn as guides to the eye, are consistent with an intrinsic Jahn-Teller splitting. (Ref. 17).

However, the intensity is twenty-five times smaller than those of E1$'$ and E2$'$ and not detectable.

Finally we mention that the ratio (γ_2/γ_3) of two of the Luttinger parameters can be obtained measuring the relative intensities of the Raman lines with $\boldsymbol{B}$ along a $< 110 >$ direction.

The intensity ratios $(I_3/I_4) : (I_{3'}/I_{4'})$ and $(I_2/I_1) : (I_{2'}/I_{1'})$ are *temperature independent* because the primed and the corresponding unprimed line have the *same initial state* and hence the thermal factors cancel. The calculated ratios are 0.7 and 1.2 for B = 6 T and $\gamma_2 \sim 0.1\gamma_3$. From the experimental ratios one can directly deduce (γ_2/γ_3) using the theoretically calculated Raman amplitudes as linear combinations of γ_2 and γ_3. The theoretical value of $(I_3/I_4) : (I_{3'}/I_{4'})$.

$$\left(\frac{3\gamma_3 - \gamma_2}{3(\gamma_3 + \gamma_2)} \frac{g_+ h_+}{f_- k}\right)^2 .$$

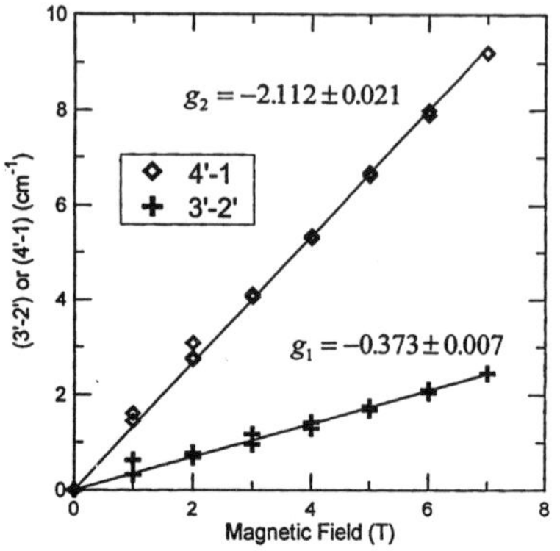

Figure 8.4 The magnetic field dependence of the spacings between lines $3'$ and $2'$ as well as that between $4'$ and 1. The straight lines passing through data for $\boldsymbol{B} \parallel [001]$ and $[111]$ permit a determination of the g-factors g_1 and g_2. (Ref. 17). From the values of g_1 and g_2 we obtain $g_{3/2} = -0.95$ and $g_{1/2} = 0.21$.

From the values of the mixing parameters g_+, h_+, f_-, and k calculated for $B =$ 6T and equating the above ratio to its experimental value of 0.65, we deduce that

$$\frac{\gamma_2}{\gamma_3} \cong 0.08.$$

Similarly the ratio $(I_2/I_1) : (I_{2'}/I_{1'})$ is

$$\left(\frac{3\gamma_3 - \gamma_2}{3(\gamma_3 + \gamma_2)} \frac{g_- h_-}{f_+ k}\right)^2.$$

Equating this to the experimental value 1.23, we find

$$\frac{\gamma_2}{\gamma_3} = 0.09.$$

For details of the scattering geometries we refer the reader to Ref. 17.

Chapter 9

Light Scattering and Fluctuations

We have seen that the intensity of the scattering of light by a fluid is intimately connected with the fluctuations in its density. We wish to describe now this qualitative statement in mathematical form and give a theory for the phenomenon of critical opalescence. We start with a few concepts in statistical physics.

9.1 The radial distribution function.

Consider a system of molecules of mass m described by the Hamiltonian function

$$H = K + \Phi = \sum_i \frac{p_i^2}{2m} + \Phi \ , \tag{9.1}$$

where K is the kinetic energy and Φ the potential energy which we suppose a function of the positions of all the molecules. Let there be N such molecules and assume their internal degrees of freedom are not excited at the temperatures of interest. A monatomic gas or liquid satisfies this requirement. We make this approximation to simplify our discussion but it is not really necessary. The probability that the system be in the volume element $d\boldsymbol{r}_1 d\boldsymbol{r}_2 \ldots d\boldsymbol{r}_N \ d\boldsymbol{p}_1 d\boldsymbol{p}_2 \ldots d\boldsymbol{p}_N$ in phase space at temperature T is

$$\frac{\exp(-\beta K - \beta\Phi) d\boldsymbol{r}_1 d\boldsymbol{r}_2 \ldots d\boldsymbol{r}_N d\boldsymbol{p}_1 d\boldsymbol{p}_2 \ldots d\boldsymbol{p}_N}{\int e^{-\beta K} d\boldsymbol{p}_1 \ldots d\boldsymbol{p}_N \int e^{-\beta\Phi} d\boldsymbol{r}_1 d\boldsymbol{r}_2 \ldots d\boldsymbol{r}_N}$$

where $\beta = (k_B T)^{-1}$. The probability that the molecules be within $d\boldsymbol{r}_1 d\boldsymbol{r}_2 \ldots d\boldsymbol{r}_N$ at $(\boldsymbol{r}_1 \boldsymbol{r}_2 \ldots \boldsymbol{r}_N)$ without regard for the values of their momenta is

$$d\boldsymbol{r}_1 \ldots d\boldsymbol{r}_N P_N(\boldsymbol{r}_1 \boldsymbol{r}_2 \ldots \boldsymbol{r}_N) = \frac{e^{-\beta\Phi} d\boldsymbol{r}_1 d\boldsymbol{r}_2 \ldots d\boldsymbol{r}_N}{\int e^{-\beta\Phi} d\boldsymbol{r}^{(N)}} \ . \tag{9.2}$$

In Eq. (9.2) we write $dr^{(N)}$ instead of $d\boldsymbol{r}_1 \dots d\boldsymbol{r}_N$ to simplify the writing. The probability of finding molecule 1 within $d\boldsymbol{r}_1$ and molecule 2 within $d\boldsymbol{r}_2$ is

$$P_2(\boldsymbol{r}_1, \boldsymbol{r}_2) d\boldsymbol{r}_1 d\boldsymbol{r}_2 = d\boldsymbol{r}_1 d\boldsymbol{r}_2 \frac{\int e^{-\beta\Phi} dr^{(N-2)}}{\int e^{-\beta\Phi} dr^{(N)}} , \tag{9.3}$$

where the integral over $dr^{(N-2)}$ is carried out over all molecular coordinates but $\boldsymbol{r}_1$ and $\boldsymbol{r}_2$. The probability of finding molecule 1 within $d\boldsymbol{r}_1$ is obviously

$$\frac{d\boldsymbol{r}_1}{V}$$

where V is the volume of the system. Now we define a function $g(r_{12})$ as follows. Given that molecule 1 is at $\boldsymbol{r}_1$, the probability of finding molecule 2 within the volume element $d\boldsymbol{r}_2$ at $\boldsymbol{r}_2$ is

$$\frac{d\boldsymbol{r}_2}{V} g(r_{12}) ,$$

with $r_{12} = |\boldsymbol{r}_1 - \boldsymbol{r}_2|$. We can relate this to P_2 as follows

$$\frac{d\boldsymbol{r}_1}{V} \frac{d\boldsymbol{r}_2}{V} g(r_{12}) = d\boldsymbol{r}_1 d\boldsymbol{r}_2 P_2(\boldsymbol{r}_1, \boldsymbol{r}_2)$$

so that

$$g(r_{12}) = V^2 P_1(\boldsymbol{r}_1, \boldsymbol{r}_2) = V^2 \frac{\int e^{-\beta\Phi} dr^{(N-2)}}{\int e^{-\beta\Phi} dr^{(N)}} . \tag{9.4}$$

Since for large values of r_{12}, the probability of finding 2 within $d\boldsymbol{r}_2$ becomes $d\boldsymbol{r}_2/V$, independent of the position of 1, $g(r_{12})$ approaches unity as $r_{12} \to \infty$. Since molecules are supposed to be unable to occupy simultaneously the same position in space $g(r_{12}) \to 0$ as $r_{12} \to 0$. Thus $g(r_{12})$ behaves as in Fig. 9.1. $g(r_{12})$ is approximately zero in a region r_{12} of the order of a molecular diameter for a dense liquid.

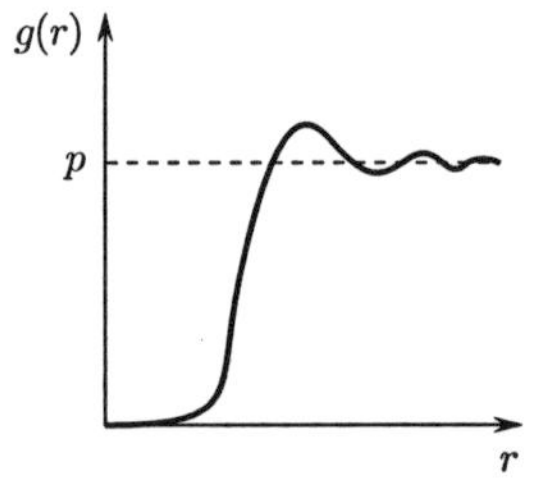

Figure 9.1 Schematic behavior of the pair distribution function for a liquid.

We now give an interpretation of $g(r_{12})$ in terms of the fluctuations in density. Let V_a be a small volume within V which we will, however, suppose large enough so that it contains a large number of molecules. We wish to investigate the fluctuations in N_a, the number of molecules within V_a. Now clearly

$$\overline{N}_a = V_a \frac{N}{V} = nV_a \ , \tag{9.5}$$

where $n = N/V$ is the average density of molecules. We define now a function $f_a(\boldsymbol{r})$ by

$$f_a(\boldsymbol{r}) = \begin{cases} 1 \text{ if } \boldsymbol{r} \text{ is within } V_a \\ 0 \text{ otherwise} \ . \end{cases} \tag{9.6}$$

Then, clearly

$$N_a = \sum_{i=1}^{N} f_a(\boldsymbol{r}_i) \ . \tag{9.7}$$

We wish to calculate

$$\overline{(N_a - \overline{N}_a)^2} = \overline{N_a^2} - \overline{N}_a^2 \ . \tag{9.8}$$

We already know $\overline{N}_a$. Now

$$\begin{aligned}
\overline{N_a^2} &= \sum_{ij} \overline{f_a(\boldsymbol{r}_i) f_a(\boldsymbol{r}_j)} = \sum_i \overline{f_a(\boldsymbol{r}_i)} + \sum_{i \neq j} \overline{f_a(\boldsymbol{r}_i) f_a(\boldsymbol{r}_j)} \\
&= \overline{N}_a + N(N-1) \overline{f_a(\boldsymbol{r}_1) f_a(\boldsymbol{r}_2)} \\
&= \overline{N}_a + N(N-1) \int d\boldsymbol{r}_1 d\boldsymbol{r}_2 f_a(\boldsymbol{r}_1) f_a(\boldsymbol{r}_2) \frac{\int e^{-\beta\Phi} d\boldsymbol{r}^{(N-2)}}{\int e^{-\beta\Phi} d\boldsymbol{r}^{(N)}} \\
&= \overline{N}_a + \frac{N(N-1)}{V^2} \int d\boldsymbol{r}_1 d\boldsymbol{r}_2 f_a(\boldsymbol{r}_1) f_a(\boldsymbol{r}_2) g(r_{12}) \\
&\approx \overline{N}_a + n^2 \int_{V_a} d\boldsymbol{r}_1 \int_{V_a} d\boldsymbol{r}_2 g(r_{12}) \\
&= \overline{N}_a + \overline{N}_a^2 + n^2 \int_{V_a} d\boldsymbol{r}_1 \int_{V_a} d\boldsymbol{r}_2 [g(r_{12}) - 1] \ .
\end{aligned}$$

If $V_a/d^3 \gg 1$, d being a molecular diameter, we can make a change of variables to $\boldsymbol{r}_{12} = \boldsymbol{r}_1 - \boldsymbol{r}_2$ and $\boldsymbol{R} = (\boldsymbol{r}_1 + \boldsymbol{r}_2)/2$. The integration over $\boldsymbol{R}$ gives simply V_a and we have $[g(r_{12}) - 1 \approx 0$ if $r_{12} > d]$

$$\overline{N_a^2} - \overline{N}_a^2 = \overline{N}_a + \overline{N}_a n \int d\boldsymbol{r}_{12} [g(r_{12}) - 1] \ ,$$

or

$$\frac{\overline{(\Delta N_a)^2}}{\overline{N}_a} = 1 + n \int d\boldsymbol{r}_{12} [g(r_{12}) - 1] = 1 + n \int d\boldsymbol{r} [g(r) - 1] \ . \tag{9.9}$$

We now consider the correlation in the numbers of molecules N_a, N_b of two separate volumes V_a, V_b. Let $\boldsymbol{r}_a$ and $\boldsymbol{r}_b$ be the positions of the centers of V_a and V_b, respectively. Since there is no overlap between V_a and V_b, $f_a(\boldsymbol{r}_i) f_b(\boldsymbol{r}_i) \equiv 0$,

$$\overline{N_a N_b} = \sum_{ij} \overline{f_a(\boldsymbol{r}_i) f_b(\boldsymbol{r}_j)} = \sum_i \overline{f_a(\boldsymbol{r}_i) f_b(\boldsymbol{r}_i)} + \sum_{i \neq j} \overline{f_a(\boldsymbol{r}_i) f_b(\boldsymbol{r}_j)} \ .$$

The first term on the right hand side clearly vanishes since a molecule cannot be simultaneously in V_a and V_b. Thus

$$\begin{aligned}\overline{N_a N_b} &= N(N-1)\int d\boldsymbol{r}_1 \int d\boldsymbol{r}_2 f_a(\boldsymbol{r}_1) f_b(\boldsymbol{r}_2)\frac{g(r_{12})}{V^2} \\ &= n^2 \int_{V_a} d\boldsymbol{r}_1 \int_{V_b} d\boldsymbol{r}_2 g(r_{12}) \approx n^2 g(r_{ab}) V_a V_b = \overline{N}_a \overline{N}_b g(r_{ab}) \ .\end{aligned}$$

Therefore

$$g(r_{ab}) = \frac{\overline{N_a N_b}}{\overline{N}_a \overline{N}_b} \ , \tag{9.10}$$

and

$$g(r_{ab}) - 1 = \frac{\overline{N_a N_b} - \overline{N}_a \overline{N}_b}{\overline{N}_a \overline{N}_b} = \frac{\overline{(N_a - \overline{N}_a)(N_b - \overline{N}_b)}}{\overline{N}_a \overline{N}_b} \ . \tag{9.11}$$

We can relate these quantities to thermodynamic variables. For this purpose we use the grand canonical ensemble applied to the volume V_a. From the theory of the grand canonical ensemble $\left(\beta = (k_B T)^{-1}\right)$

$$\overline{(N_a - \overline{N}_a)^2} = \frac{1}{\beta}\left(\frac{\partial \overline{N}_a}{\partial \mu}\right)_{V_a, T} \ , \tag{9.12}$$

where μ is the chemical potential of the molecules, namely

$$\mu = \frac{G}{N} = \frac{F + PV}{N} = f + Pv \ ; \tag{9.13}$$

In this equation G is the Gibbs free energy, F the Helmholtz free energy, and f and v the free energy and volume per molecule. Now

$$\begin{aligned}\left(\frac{\partial \mu}{\partial \overline{N}_a}\right)_{V_a, T} &= \frac{dv}{d\overline{N}_a}\frac{\partial}{\partial v}(f + Pv) \\ &= \frac{dv}{d\overline{N}_a}\left[\left(\frac{\partial f}{\partial v}\right)_T + \left(\frac{\partial P}{\partial v}\right)_T v + P\right] = -\frac{v^2}{\overline{N}_a}\left(\frac{\partial P}{\partial v}\right)_T \ ,\end{aligned} \tag{9.14}$$

since

$$P = -\left(\frac{\partial f}{\partial v}\right)_T \ . \tag{9.15}$$

Therefore

$$\frac{\overline{(N_a - \overline{N}_a)^2}}{\overline{N}_a} = -\frac{k_B T}{v^2}\left(\frac{\partial v}{\partial P}\right)_T = k_B T n \kappa_T \ , \tag{9.16}$$

where

$$\kappa_T = -\frac{1}{v}\left(\frac{\partial v}{\partial P}\right)_T \tag{9.17}$$

is the isothermal compressibility.

Fig. 9.2 shows schematically the isothermal $p - v$ curves of a fluid above and below the critical temperature T_c. We see that for $T \leq T_c$, $(\partial v/\partial P)_T$ becomes very large so that the fluctuations in the density become large.

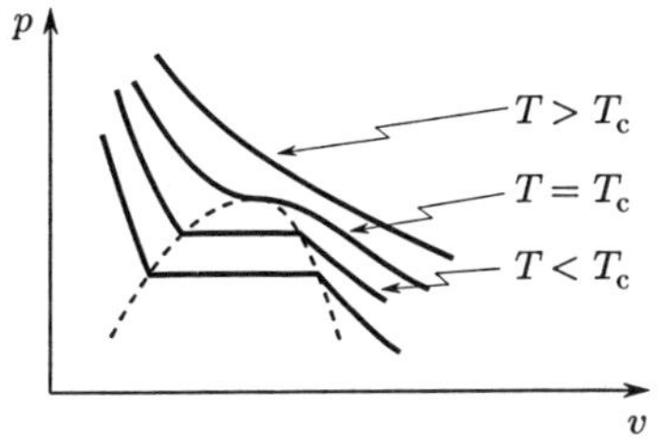

Figure 9.2 Schematic shape of the isothermal curves of a fluid near the critical temperature T_c.

9.2 The theory of critical opalescence.

We consider the scattering of light by a system composed of N molecules at positions $\boldsymbol{r}_1, \boldsymbol{r}_2, \dots \boldsymbol{r}_N$. Let $f(\theta)$ be the scattering amplitude for a single molecule, θ being the scattering angle. If $\boldsymbol{k}$ is the incident and $\boldsymbol{k}'$ the scattered wave vector, the optical path difference for scattering from a molecule at O and another at $\boldsymbol{r}_i$ is $(\boldsymbol{k}-\boldsymbol{k}')\cdot\boldsymbol{r}_i$. The scattering amplitude for all molecules is, thus,

$$\sum_{i=1}^{N} \exp(i(\boldsymbol{k}-\boldsymbol{k}')\cdot\boldsymbol{r}_i) f(\theta) = f(\theta)\sum_{i=1}^{N}\exp(i\boldsymbol{q}\cdot\boldsymbol{r}_i)$$

where

$$\boldsymbol{q} = \boldsymbol{k} - \boldsymbol{k}'$$

is the momentum transferred (in units of $\hbar$) in the scattering process. The scattered wave amplitude squared, or intensity, is proportional to

$$I_0 \left| \sum_{i=1}^{N} \exp(i\boldsymbol{q}\cdot\boldsymbol{r}_i) \right|^2 \tag{9.18}$$

where

$$I_0 = |f(\theta)|^2$$

is the intensity of scattered radiation for a single molecule. In this expression $\boldsymbol{r}_i$ is taken as fixed but each molecule is subjected to collisions so that the average scattered intensity is

$$\begin{aligned} I(q) &= I_0 \overline{\left| \sum_{i=1}^{N} \exp(i\boldsymbol{q}\cdot\boldsymbol{r}_i) \right|^2} \\ &= I_0 \left[N + \sum_{i\neq j} \overline{\exp(i\boldsymbol{q}\cdot(\boldsymbol{r}_i-\boldsymbol{r}_j))} \right] \\ &= I_0 \left[N + n^2 \int d\boldsymbol{r}_1 \int d\boldsymbol{r}_2 \exp(i\boldsymbol{q}\cdot(\boldsymbol{r}_1-\boldsymbol{r}_2)) g(r_{12}) \right] \\ &= I_0 \left[N + n^2 \int d\boldsymbol{r}_1 \int d\boldsymbol{r}_2 \exp(i\boldsymbol{q}\cdot(\boldsymbol{r}_1-\boldsymbol{r}_2)) \{g(r_{12})-1\} \right] \end{aligned} \tag{9.19}$$

where, assuming $\boldsymbol{q} \neq 0$, we have set $n^2 \int d\boldsymbol{r}_1 \int d\boldsymbol{r}_2 \exp(i\boldsymbol{q}\cdot(\boldsymbol{r}_1 - \boldsymbol{r}_2)) = 0$

If $|\boldsymbol{k}| \approx |\boldsymbol{k}'|$ we have

$$|\boldsymbol{q}| \cong 2|\boldsymbol{k}| \sin\frac{\theta}{2} = \frac{4\pi}{\lambda}\sin\frac{\theta}{2} \,. \tag{9.20}$$

If we take $\theta \approx 90^o$, $|\boldsymbol{q}| \sim (4\pi/\lambda) \sim 10^5$ cm^{-1} for scattering of visible radiation. Thus, if $g(r_{12}) - 1 \approx 0$ except for $0 \approx r_{12} \lesssim d$, $d \approx 10^{-8}$ cm, $\boldsymbol{q}\cdot\boldsymbol{r}_{12}$ in the region in which $g(r_{12}) - 1$ is different from zero is of the order of 10^{-3} so that we can approximate

$$\begin{aligned} I(q) &= NI_0\left[1 + n\int d\boldsymbol{r}\exp(i\boldsymbol{q}\cdot\boldsymbol{r})\{g(r) - 1\}\right] \\ &\approx NI_0\left[1 + n\int d\boldsymbol{r}\,\{g(r) - 1\}\right] \end{aligned} \tag{9.21}$$

and

$$\frac{I(q)}{NI_0} \simeq \frac{\overline{(\Delta N_a)^2}}{\overline{N}_a} \,. \tag{9.22}$$

When we are near a phase transition, the fluctuations are large and we can no longer make this approximation. In that case, experiments show that, for $\theta \neq 0$,

$$\frac{I(q)}{NI_0} = \frac{1}{A(T - T_c) + B} \,, \tag{9.23}$$

where

$$\frac{1}{B} \propto \frac{\lambda^2}{\sin^2\frac{\theta}{2}} \,. \tag{9.24}$$

It is clear that the experimental results show that $g(r) - 1$ cannot go to zero very rapidly in the vicinity of the critical point. We proceed using a method invented by Einstein and modified by Ornstein and Zernicke. Take a system in equilibrium whose partition function is

$$Z = \int e^{-\beta H} d\Gamma = e^{-\beta F} \,, \tag{9.25}$$

where the integration extends over the whole phase space. Suppose now that we impose a restriction C upon the system, the probability that that restriction is realized is

$$P_C = \frac{1}{Z}\int_C e^{-\beta H} d\Gamma \tag{9.26}$$

where now the integration over Γ extends only over those regions in phase space consistent with the restriction C. Now, if F_c is the free energy of the system when it is subject to C

$$\int_C e^{-\beta H} d\Gamma = e^{-\beta F_c} \,, \tag{9.27}$$

and

$$P_C = e^{-\beta(F_c - F)} \,. \tag{9.28}$$

We imagine now that the restriction C is such that the density of particles in the system is not uniformly equal to n but is a function of position $n(\boldsymbol{r})$ which differs by a small quantity $\delta n(\boldsymbol{r})$ from n, *i.e.*,

$$n(\boldsymbol{r}) = n + \delta n(\boldsymbol{r}) \tag{9.29}$$

According to Eq. (9.28), the probability of this occurring is

$$P\{n(\boldsymbol{r})\} = \exp[-\beta(F\{n(\boldsymbol{r})\} - F)] \tag{9.30}$$

Here $F\{n(\boldsymbol{r})\}$, the Helmholtz free energy of the system subject to the requirement that the density of particles be $n(\boldsymbol{r})$, is a functional of $n(\boldsymbol{r})$, *i.e.*, it depends on the choice of the function $n(\boldsymbol{r})$. F is, of course, the equilibrium free energy. Let $\varphi(\boldsymbol{r})$ be the density of free energy at $\boldsymbol{r}$. Then

$$F\{n(\boldsymbol{r})\} = \int d\boldsymbol{r}\varphi(\boldsymbol{r}) \ . \tag{9.31}$$

We can write, for small $\delta n(\boldsymbol{r})$,

$$\varphi(\boldsymbol{r}) = \overline{\varphi} + \frac{a}{2}\,[\delta n(\boldsymbol{r})]^2 + \frac{b}{2}\,[\nabla n(\boldsymbol{r})]^2 + \dots \ . \tag{9.32}$$

This is an expansion of $\varphi(\boldsymbol{r})$ assumed to be a function of $\delta n(\boldsymbol{r})$ and its gradient

$$\nabla\delta n(\boldsymbol{r}) = \nabla n(\boldsymbol{r}) \ .$$

$\overline{\varphi}$ is the equilibrium free energy per unit volume. The coefficients of $\delta n(\boldsymbol{r})$ and $\nabla\delta n(\boldsymbol{r})$ in Eq. (9.32) must vanish because $\overline{\varphi}$ must be a minimum of $\varphi(\boldsymbol{r})$. For the same reason a and b must be positive. We can describe the fluctuation $\delta n(\boldsymbol{r})$ in terms of its Fourier components which we may regard as generalized coordinates:

$$\delta n(\boldsymbol{r}) = \sum_q \delta n_q \exp(i\boldsymbol{q}\cdot\boldsymbol{r}) \tag{9.33}$$

and

$$\delta n_q = \frac{1}{V}\int d\boldsymbol{r}\exp(-i\boldsymbol{q}\cdot\boldsymbol{r})\delta n(\boldsymbol{r}) \ . \tag{9.34}$$

Since $\delta n(\boldsymbol{r})$ is real, we must have

$$\delta n_{-q} = \delta n_q^* \ . \tag{9.35}$$

Now, the physical meaning of a is

$$a = \left(\frac{\partial^2\varphi}{\partial n^2}\right)_T = \left(\frac{\partial\mu}{\partial n}\right)_T \ . \tag{9.36}$$

By virtue of Eq. (8.14) $a = 0$ for $T \leqq T_c$ during a phase transition. Now

$$\delta F = F\{n(\boldsymbol{r})\} - F = \frac{a}{2}\int d\boldsymbol{r}[\delta n(\boldsymbol{r})]^2 + \frac{b}{2}\int d\boldsymbol{r}[\nabla\delta n(\boldsymbol{r})]^2 \ ,$$

or

$$\delta F = \frac{V}{2}\sum_q |\delta n_q|^2(a + bq^2) \ . \tag{9.37}$$

Therefore

$$P\{n(\boldsymbol{r})\} = \exp\left(-\frac{\beta V}{2}\sum_q |\delta n_q|^2(a+bq^2)\right) . \tag{9.38}$$

Because of Eq. (9.35) we cannot regard all δn_q as independent coordinates. Since $|\delta n_{-q}|^2 = |\delta n_q|^2$ it is enough to restrict the sum over $\boldsymbol{q}$ to half its range, i.e., in such a way that, if the sum contains $\boldsymbol{q}$, then it does not contain $-\boldsymbol{q}$. This can be done, for example, by summing over all $\boldsymbol{q}$'s such that $q_x \geq 0$. Let α_q and β_q be the real and imaginary parts of δn_q:

$$\delta n_q = \alpha_q + i\beta_q .$$

With the restriction $q_x \geq 0$, all α_q and β_q are independent and, from Eq. (9.38), by virtue of the equipartition theorem

$$\overline{|\delta n_q|^2} = \overline{\alpha_q^2} + \overline{\beta_q^2} = \frac{k_B T}{V(a+bq^2)} . \tag{9.39}$$

From Eq. (9.11)

$$g(|\boldsymbol{r}-\boldsymbol{r}'|) - 1 = \frac{\overline{\delta n(\boldsymbol{r})\delta n(\boldsymbol{r}')}}{n^2} . \tag{9.40}$$

But

$$\begin{aligned}\overline{\delta n(\boldsymbol{r})\delta n(\boldsymbol{r}')} &= \overline{\sum_{qq'} \delta n_q \delta n^*_{q'}} \exp(i\boldsymbol{q}\cdot\boldsymbol{r} - i\boldsymbol{q}'\cdot\boldsymbol{r}') \\ &= \sum_{qq'} \exp(i\boldsymbol{q}\cdot\boldsymbol{r} - i\boldsymbol{q}'\cdot\boldsymbol{r}')\left(\overline{\alpha_q\alpha_{q'}} + \overline{\beta_q\beta_{q'}} - \overline{i\alpha_q\beta_{q'}} + \overline{i\beta_q\alpha_{q'}}\right) .\end{aligned}$$

Using $\alpha_{-q} = \alpha_q$ and $\beta_{-q} = -\beta_q$ which follow from Eq. (9.35) and the fact that the other coordinates are independent the sum over $\boldsymbol{q}$ and $\boldsymbol{q}'$ can be simplified. Because of Eq. (9.39)

$$\overline{\delta n(\boldsymbol{r})\delta n(\boldsymbol{r}')} = \sum_q \frac{k_B T}{V(a+bq^2)} \exp(i\boldsymbol{q}\cdot(\boldsymbol{r}-\boldsymbol{r}')) . \tag{9.41}$$

Thus

$$g(r) - 1 = \frac{k_B T}{(2\pi)^3 n^2}\int \frac{d\boldsymbol{q}\exp(i\boldsymbol{q}\cdot\boldsymbol{r})}{a+bq^2} , \tag{9.42}$$

or

$$g(r) - 1 = \frac{k_B T}{4\pi n^2 b r}\exp\left(-\left(\frac{a}{b}\right)^{1/2} r\right) . \tag{9.43}$$

We note that at the critical point ($a = 0$) $g(r) - 1$ is a very slowly decreasing function of r. Now, from Eqs. (9.21) and (9.42)

$$\begin{aligned}\frac{I(q)}{NI_0} &= 1 + n\int [g(r)-1]\exp(-i\boldsymbol{q}\cdot\boldsymbol{r})d\boldsymbol{r} , \\ &= 1 + \frac{k_B T}{n(a+bq^2)} .\end{aligned} \tag{9.44}$$

For T near T_c, the second term predominates and $a = 0$ so that

$$\frac{I(q)}{NI_0} \approx \frac{k_B T}{nbq^2} = \frac{k_B T}{16\pi^2 nb} \frac{\lambda^2}{\sin^2 \frac{\theta}{2}} , \tag{9.45}$$

where use was made of Eq. (9.20). We note that the main feature of this result is in agreement with the experimental observation mentioned in Eqs. (9.23) and (9.24).

Appendix I

A.1 Description of the State of Polarization of Light

A plane monochromatic electromagnetic wave propagating along the direction of the unit vector $\hat{n}$ can be described by its electric field $\boldsymbol{E}(\boldsymbol{r},t)$ and magnetic induction $\boldsymbol{B}(\boldsymbol{r},t)$ given by

$$\boldsymbol{E}(\boldsymbol{r},t) = \Re\left[\boldsymbol{E}_0 \exp(i(\boldsymbol{k}\cdot\boldsymbol{r} - \omega t))\right] \tag{1}$$

and

$$\boldsymbol{B}(\boldsymbol{r},t) = \hat{\boldsymbol{n}} \times \boldsymbol{E}(\boldsymbol{r},t) \tag{2}$$

where

$$\boldsymbol{k} = \frac{2\pi}{\lambda}\hat{\boldsymbol{n}} \ , \tag{3}$$

λ is the wavelength of the electromagnetic wave and $\boldsymbol{E}_0$ is a complex vector perpendicular to $\hat{\boldsymbol{n}}$. Selecting a set of unit vectors $\hat{\mathbf{e}}_1, \hat{\mathbf{e}}_2$ in the plane of $\boldsymbol{E}$ so that $\hat{\mathbf{e}}_1$, $\hat{\mathbf{e}}_2$ and $\hat{\boldsymbol{n}}$ form a right–handed triad of orthonormal vectors we can write

$$\hat{\mathbf{e}}_1 \cdot \boldsymbol{E}_0 = a_1 e^{-i\delta_1} \tag{4}$$

and

$$\hat{\mathbf{e}}_2 \cdot \boldsymbol{E}_0 = a_2 e^{-i\delta_2} \tag{5}$$

where a_1, a_2, δ_1 and δ_2 are real quantities and a_1 and a_2 are taken to be positive without loss of generality. We select the axes $\hat{\mathbf{e}}_1$, $\hat{\mathbf{e}}_2$ so that $a_1 > a_2$. The case in which $a_1 = a_2$ will be discussed separately. We can choose the origin of time so that $\delta_1 = 0$; only the phase difference

$$\delta = \delta_2 - \delta_1$$

is physically significant. Thus we can write

$$\boldsymbol{E}_0 = a_1\hat{\mathbf{e}}_1 + a_2\hat{\mathbf{e}}_2 e^{-i\delta} \ . \tag{6}$$

The three parameters a_1, a_2 and δ completely specify the state of polarization of the wave.

On a fixed plane, say $z = 0$, the components of the electric field are

$$E_1 = a_1 \cos \omega t \tag{7}$$

$$E_2 = a_2 \cos(\omega t + \delta) . \tag{8}$$

Eliminating the time we obtain

$$\left(\frac{E_1}{a_1}\right)^2 + \left(\frac{E_2}{a_2}\right)^2 - \frac{E_1 E_2 \cos \delta}{a_1 a_2} = \sin^2 \delta , \tag{9}$$

i.e., the tip of the vector $\boldsymbol{E}$ traces an ellipse in the $z = 0$ plane (and, of course similar, but rotated, ellipses in all other planes). This ellipse is not referred to its principal axes. We can, however, perform a transformation to new axes $\hat{\mathbf{e}}_1'$ and $\hat{\mathbf{e}}_2'$ forming an angle ψ with $\hat{\mathbf{e}}_1$ and $\hat{\mathbf{e}}_2$, respectively. The transformation is

$$\left.\begin{aligned} \hat{\mathbf{e}}_1 &= \hat{\mathbf{e}}_1' \cos \psi - \hat{\mathbf{e}}_2' \sin \psi \\ \hat{\mathbf{e}}_2 &= \hat{\mathbf{e}}_1' \sin \psi + \hat{\mathbf{e}}_2' \cos \psi \end{aligned}\right\} . \tag{10}$$

The complex vector $\boldsymbol{E}_0$ expressed in terms of $\hat{\mathbf{e}}_1'$ and $\hat{\mathbf{e}}_2'$ reads

$$\boldsymbol{E}_0 = \left(a_1 \cos \psi + a_2 e^{-i\delta} \sin \psi\right)\hat{\mathbf{e}}_1' + \left(-a_1 \sin \psi + a_2 e^{-i\delta} \cos \psi\right)\hat{\mathbf{e}}_2' . \tag{11}$$

We select ψ so that the components of $\boldsymbol{E}_0$ with respect to $\hat{\mathbf{e}}_1'$ and $\hat{\mathbf{e}}_2'$ differ in phase by $\pi/2$, *i.e.*, we require their ratio to be purely imaginary. This condition is

$$\frac{-a_1 \sin \psi + a_2 e^{i\delta} \cos \psi}{a_1 \cos \psi + a_2 e^{i\delta} \sin \psi} = -\frac{-a_1 \sin \psi + a_2 e^{-i\delta} \cos \psi}{a_1 \cos \psi + a_2 e^{-i\delta} \sin \psi} .$$

After some simple algebraic manipulations we obtain

$$\tan 2\psi = \frac{2a_1 a_2 \cos \delta}{a_1^2 - a_2^2} . \tag{12}$$

We now define the real numbers $a > 0$ and α by

$$a_1 \cos \psi + a_2 e^{-i\delta} \sin \psi = a e^{-i\alpha} . \tag{13}$$

Since the phase of $-a_1 \sin \psi + a_2 e^{-i\delta} \cos \psi$ differs from that of $a e^{-i\alpha}$ by $\pi/2$, a positive number b exists such that

$$-a_1 \sin \psi + a_2 e^{-i\delta} \cos \psi = \pm i b e^{-i\alpha} . \tag{14}$$

Then

$$\boldsymbol{E}_0 = \left(a\hat{\mathbf{e}}_1' \pm i b \hat{\mathbf{e}}_2'\right) e^{-i\alpha} \tag{15}$$

and

$$\boldsymbol{E}_0^* \cdot \boldsymbol{E}_0 = a^2 + b^2 = a_1^2 + a_2^2 . \tag{16}$$

Clearly $\boldsymbol{E}_0^* \cdot \boldsymbol{E}_0$ is proportional to the intensity of the wave. The phase angle α is, of course, related to the original phase difference δ. In fact, we find

$$\boldsymbol{E}_0 \cdot \boldsymbol{E}_0 = \left(a^2 - b^2\right) e^{-2i\alpha} = a_1^2 + a_2^2 e^{-2i\delta} . \tag{17}$$

Designating by E_1' and E_2' the components of $\boldsymbol{E}(0,t)$ with respect to $\hat{\mathbf{e}}_1'$ and $\hat{\mathbf{e}}_2'$ we obtain

$$\left.\begin{aligned} E_1' &= a\cos(\omega t+\alpha)\ , \\ E_2' &= \pm b\sin(\omega t+\alpha)\ . \end{aligned}\right\} \tag{18}$$

Thus, the tip of the vector $\boldsymbol{E}$ describes the ellipse

$$\left(\frac{E_1'}{a}\right)^2 + \left(\frac{E_2'}{b}\right)^2 = 1\ . \tag{19}$$

When the positive sign is taken in the second Eq. (18) the tip of the electric vector traverses its path in the positive or counterclockwise sense. For the negative sign the sense of rotation is clockwise or negative.

The quantity

$$S_0 = a^2 + b^2 = a_1^2 + a_2^2 \tag{20}$$

is proportional to the intensity of the radiation. We now define the angle χ by

$$\tan\chi = \pm\frac{b}{a}\ . \tag{21}$$

We take the + sign for positive helicity and the − sign for negative helicity. Thus, in the first case $0 < \chi < \pi/4$ while in the second $-\pi/4 < \chi < 0$. (See Fig. A.1.)

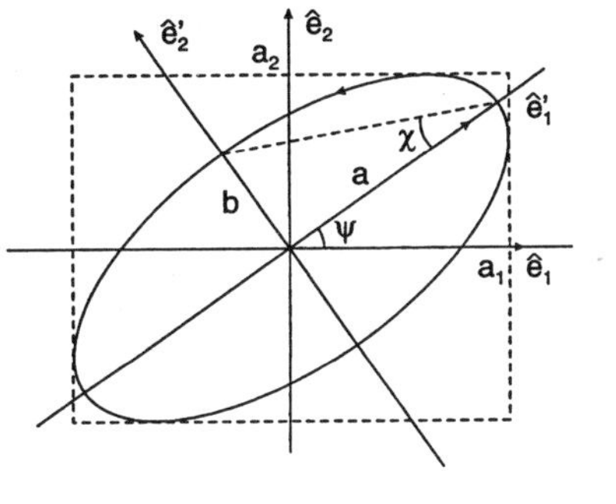

Figure A.1 Path of the tip of the electric field for a plane electromagnetic wave with arbitrary polarization.

Squaring the absolute values of Eqs. (13) and (14) and subtracting we find

$$a_1^2 - a_2^2 = \left(a^2 - b^2\right)\cos 2\psi \tag{22}$$

after use of Eq. (12). From Eq. (21) we obtain

$$\cos 2\chi = \frac{a^2-b^2}{a^2+b^2}\ , \qquad \sin 2\chi = \pm\frac{2ab}{a^2+b^2}\ . \tag{23}$$

Multiplying Eq. (13) with the complex conjugate of Eq. (14) and separating the imaginary part we find

$$a_1 a_2 \sin\delta = \mp ab\ . \tag{24}$$

We define

$$\left.\begin{aligned} S_1 &= a_1^2 - a_2^2 = S_0 \cos 2\psi \cos 2\chi \ , \\ S_2 &= 2a_1 a_2 \cos\delta = S_0 \sin 2\psi \cos 2\chi \ , \\ S_3 &= -2a_1 a_2 \sin\delta = S_0 \sin 2\chi \ . \end{aligned}\right\} \tag{25}$$

The four parameters $S_0 = a_1^2 + a_2^2$, S_1, S_2, S_3 are called the Stokes parameters. They clearly satisfy

$$S_0^2 = S_1^2 + S_2^2 + S_3^2 \ . \tag{26}$$

When $a_1 = a_2$,

$$\boldsymbol{E}_0 = a_1\left(\hat{\mathbf{e}}_1 + \hat{\mathbf{e}}_2 e^{-i\delta}\right) \tag{27}$$

and

$$\left.\begin{aligned} E_1 &= a_1 \cos\omega t \ , \\ E_2 &= a_1 \cos(\omega t + \delta) \ . \end{aligned}\right\} \tag{28}$$

Transforming to the new axes defined by Eqs. (10) we obtain after setting $\psi = \pi/4$

$$\boldsymbol{E}_0 = \sqrt{2} a_1 e^{-i\delta/2} \cos\frac{\delta}{2} \hat{\mathbf{e}}_1' - i\sqrt{2} a_1 e^{-i\delta/2} \sin\frac{\delta}{2} \hat{\mathbf{e}}_2' \tag{29}$$

or

$$E_1' = \sqrt{2} a_1 \cos\frac{\delta}{2} \cos\left(\omega t + \tfrac{1}{2}\,\delta\right) \ , \tag{30}$$

and

$$E_2' = -\sqrt{2} a_1 \sin\frac{\delta}{2} \sin\left(\omega t + \tfrac{1}{2}\,\delta\right) \ .$$

When $a_1 = a_2$, $2\psi = \pi/2$ and $S_1 = 0$.

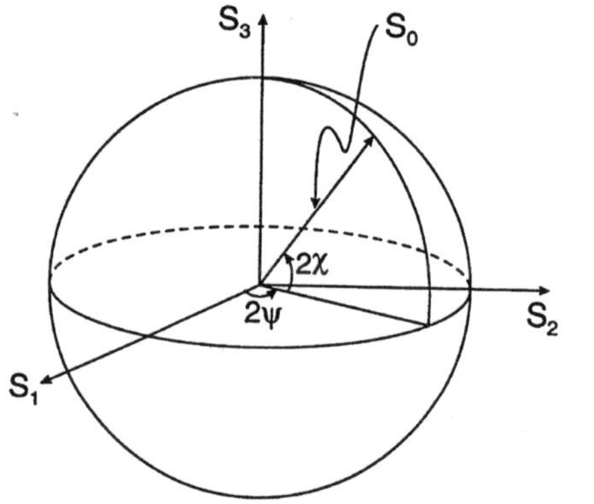

Figure A.2 The Stokes parameters represented in the Poincaré sphere.

The state of polarization of a monochromatic wave can be represented by a vector of length S_0 and components S_1, S_2, S_3. The different tips of these vectors lie in a sphere called the Poincaré sphere. (See Fig. A.2.) Plane polarized radiation corresponds to $\delta = n\pi$ where n is an integer. These cases are shown in Fig. A.3 for n even and odd. Circularly polarized radiation is characterized by

$$\delta = 2n\pi \mp \frac{\pi}{2} \ .$$

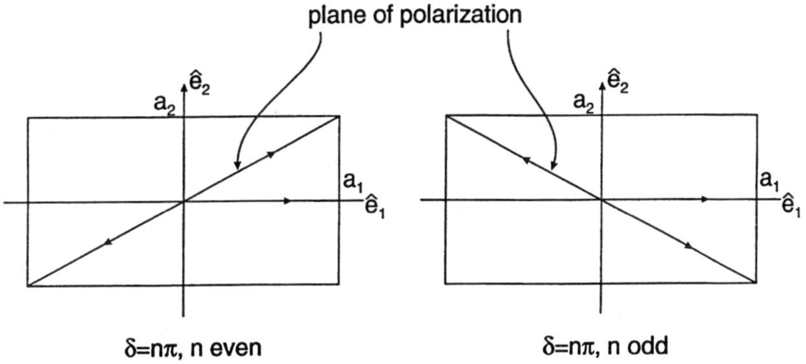

Figure A.3 Schematic representation of the electric vector in plane polarized waves.

From

$$\cos\left(\omega t + 2n\pi \mp \frac{\pi}{2}\right) = \pm \sin \omega t$$

we conclude that when $\delta = -\pi/2$ we have left–hand circularly polarized light (positive helicity) while for $\delta = \pi/2$ we have right–hand circularly polarized light.

Linear polarization is characterized by $\chi = 0$, *i.e.*, by the points on the equator of the Poincaré sphere. Circularly polarized radiation with positive helicity corresponds to $\chi = \pi/4$ (ψ arbitrary), *i.e.*, to the north pole of the Poincaré sphere. Similarly a circularly polarized wave with negative helicity is represented by the south pole $(0, 0, -S_0)$.

For circularly polarized radiation

$$\boldsymbol{E}_0 = a_1 \hat{\boldsymbol{e}}_1 + a_1 \hat{\boldsymbol{e}}_2 e^{\pm \frac{i\pi}{2}} = a_1 \sqrt{2} \hat{\boldsymbol{e}}_\pm \tag{31}$$

where

$$\hat{\boldsymbol{e}}_\pm = \frac{1}{\sqrt{2}} (\hat{\boldsymbol{e}}_1 \pm i \hat{\boldsymbol{e}}_2) \; . \tag{32}$$

The upper (lower) sign corresponds to positive (negative) helicity. The following properties of $\hat{\boldsymbol{e}}_\pm$ follow immediately

$$\begin{aligned} \hat{\boldsymbol{e}}_\pm^* &= \hat{\boldsymbol{e}}_\mp \; , \\ \hat{\boldsymbol{e}}_\pm \cdot \hat{\boldsymbol{e}}_\pm &= 0 \; , \\ \hat{\boldsymbol{e}}_\pm^* \cdot \hat{\boldsymbol{e}}_\pm &= 1 \; . \end{aligned} \tag{33}$$

We note that for circularly polarized radiation

$$E_{02} = \pm i E_{01} \tag{34}$$

where the $+(-)$ sign corresponds to positive helicity or left–hand polarization (negative helicity or right–hand polarization).

The Poynting vector of the wave is

$$\boldsymbol{S} = \frac{c}{4\pi}\boldsymbol{E}\times\boldsymbol{B} = \frac{cE^2}{4\pi}\hat{\boldsymbol{n}}\ . \tag{35}$$

Using

$$\boldsymbol{E}(\boldsymbol{r},t) = \tfrac{1}{2}\ \boldsymbol{E}_0 \exp(i(\boldsymbol{k}\cdot\boldsymbol{r}-\omega t)) + \tfrac{1}{2}\ \boldsymbol{E}_0^* \exp(-i(\boldsymbol{k}\cdot\boldsymbol{r}-\omega t))$$

we find

$$\boldsymbol{S}(\boldsymbol{r},t) = \frac{c\hat{\boldsymbol{n}}}{8\pi}\Big(\boldsymbol{E}_0\cdot\boldsymbol{E}_0^* + \frac{1}{2}\boldsymbol{E}_0\cdot\boldsymbol{E}_0\exp(2i(\boldsymbol{k}\cdot\boldsymbol{r}-\omega t)) + \frac{1}{2}\boldsymbol{E}_0^*\cdot\boldsymbol{E}_0^*\exp(-2i(\boldsymbol{k}\cdot\boldsymbol{r}-\omega t))\Big)\ .$$

The average over a period is

$$\langle \boldsymbol{S}(\boldsymbol{r},t)\rangle = \frac{c\hat{\boldsymbol{n}}\boldsymbol{E}_0\cdot\boldsymbol{E}_0^*}{8\pi}\ . \tag{36}$$

In general, the time average of the scalar, vector or tensor product of two sinusoidal fields

$$\boldsymbol{u} = \Re\big(\boldsymbol{u}_0 e^{-i\omega t}\big)\ ,\quad \boldsymbol{v} = \Re\big(\boldsymbol{v}_0 e^{-i\omega t}\big)$$

is

$$\langle \boldsymbol{u}\cdot\boldsymbol{v}\rangle = \tfrac{1}{4}\ (\boldsymbol{u}_0\cdot\boldsymbol{v}_0^* + \boldsymbol{u}_0^*\cdot\boldsymbol{v}_0) = \tfrac{1}{2}\ \Re(\boldsymbol{u}_0\cdot\boldsymbol{v}_0^*)\ , \tag{37}$$

for the scalar product, while for the vector product it is

$$\langle \boldsymbol{u}\times\boldsymbol{v}\rangle = \tfrac{1}{4}\ (\boldsymbol{u}_0\times\boldsymbol{v}_0^* + \boldsymbol{u}_0^*\times\boldsymbol{v}_0)\ .$$

So far we have discussed only monochromatic waves which, by definition, are polarized, their polarization being characterized by the Stokes parameters S_0, S_1, S_2, S_3 satisfying Eq. (26). We turn now to the study of partially polarized light.

In general the electric and magnetic fields of an electromagnetic wave are not monochromatic. We shall consider here a plane wave having a power spectrum of relatively narrow width $\Delta\omega$ centered at ω. Then, on a fixed plane normal to the direction of propagation $\hat{\boldsymbol{n}}$ (say $\boldsymbol{r}\cdot\hat{\boldsymbol{n}} = 0$), the electric field $\boldsymbol{E}$ can be represented by

$$\boldsymbol{E} = \Re\big[\boldsymbol{E}_0(t)e^{-i\omega t}\big] \tag{38}$$

where $\boldsymbol{E}_0(t)$ is a slowly varying complex vector normal to $\hat{\boldsymbol{n}}$. We can express $\boldsymbol{E}_0(t)$ in the form

$$\boldsymbol{E}_0(t) = a_1(t)\hat{e}_1 e^{-i\delta_1(t)} + a_2(t)\hat{e}_2 e^{-i\delta_2(t)}\ . \tag{39}$$

In analogy with the definitions of S_0, S_1, S_2, S_3 for polarized light we now define the Stokes parameters for partially polarized light as follows:

$$S_0 = \langle a_1^2\rangle + \langle a_2^2\rangle\ ,$$

$$S_1 = \langle a_1^2\rangle - \langle a_2^2\rangle \ ,$$
$$S_2 = \langle 2a_1a_2\cos\delta\rangle \ , \tag{40}$$

and

$$S_3 = -\langle 2a_1a_2\sin\delta\rangle \ .$$

The time average of $f(t)$ is defined by

$$\langle f\rangle = \lim_{T\to\infty}\frac{1}{T}\int_0^T f(t)dt \ .$$

We consider now the two functions $a_1(t)e^{-i\delta_1(t)}$ and $a_2(t)e^{-i\delta_2(t)}$ and make use of the Schwarz inequality

$$\left|\int_0^T a_1(t)e^{i\delta_1(t)}a_2(t)e^{-i\delta_2(t)}dt\right|^2 \le \int_0^T\left|a_1(t)e^{-i\delta_1(t)}\right|^2 dt\int_0^T\left|a_2(t)e^{-i\delta_2(t)}\right|^2 dt$$

to obtain

$$\langle a_1(t)a_2(t)\cos\delta(t)\rangle^2 + \langle a_1(t)a_2(t)\sin\delta(t)\rangle^2 \le \langle a_1^2(t)\rangle\langle a_2^2(t)\rangle \ .$$

It follows immediately that

$$S_0^2 \ge S_1^2 + S_2^2 + S_3^2 \ . \tag{41}$$

The polarization characteristics of a polychromatic wave with a relatively narrow power spectrum are specified by the tensor

$$\boldsymbol{J} = \langle \boldsymbol{E}_0\boldsymbol{E}_0^*\rangle \ . \tag{42}$$

With respect to the unit vectors $\hat{\mathbf{e}}_1, \hat{\mathbf{e}}_2$, $\boldsymbol{J}$ is specified by the 2×2 Hermitian matrix

$$\boldsymbol{J} = \frac{1}{2}\begin{pmatrix} S_0+S_1 & S_2-iS_3 \\ S_2+iS_3 & S_0-S_1\end{pmatrix} = \frac{1}{2}\sum_{\mu=0}^{3} S_\mu\tau_\mu \tag{43}$$

where

$$\tau_0 = \begin{pmatrix}1&0\\0&1\end{pmatrix}, \tau_1 = \begin{pmatrix}1&0\\0&-1\end{pmatrix}, \tau_2 = \begin{pmatrix}0&1\\1&0\end{pmatrix}, \tau_3 = \begin{pmatrix}0&-i\\i&0\end{pmatrix} . \tag{44}$$

With respect to the axes $\hat{\mathbf{e}}_1', \hat{\mathbf{e}}_2'$ obtained from $\hat{\mathbf{e}}_1, \hat{\mathbf{e}}_2$ after a rotation by ψ about $\hat{\boldsymbol{n}}$, the components of $\boldsymbol{E}_0$ become

$$\begin{aligned} E_{01}' &= E_{01}\cos\psi + E_{02}\sin\psi \ , \\ E_{02}' &= -E_{01}\sin\psi + E_{02}\cos\psi \ . \end{aligned} \tag{45}$$

The Stokes parameters defined with respect to the new axes are obtained calculating $\langle E_{0i}E_{0j}^*\rangle$, $i,j = 1,2$. We find

$$\left.\begin{aligned} S_0' &= S_0 \ , \\ S_1' &= S_1\cos 2\psi + S_2\sin 2\psi \ , \\ S_2' &= -S_1\sin 2\psi + S_2\cos 2\psi \ , \\ S_3' &= S_3 \ . \end{aligned}\right\} \tag{46}$$

The quantity S_3 is a measure of the degree of circular polarization of the light. It can vary between -1 and 1. If $S_3 = 0$, we can select ψ so that $S_2 = 0$. With this particular choice of axes the components E'_{01} and E'_{02} are uncorrelated, *i.e.*,

$$\langle E'_{01} E'^{*}_{02} \rangle = 0 \ .$$

The polarization tensor $\boldsymbol{\rho}$ is defined by

$$\boldsymbol{\rho} = \boldsymbol{J}/S_0 \ . \tag{47}$$

We note that

$$Tr\boldsymbol{\rho} = 1 \tag{48}$$

and

$$\det \boldsymbol{\rho} = \rho_{11}\rho_{22} - \rho_{12}\rho_{21} = \frac{1}{4}\left(1 - \frac{S_1^2 + S_2^2 + S_3^2}{S_0^2}\right) \ . \tag{49}$$

For a polarized wave, $\det \boldsymbol{\rho} = 0$ while for unpolarized or natural light $\langle E_{0i} E^{*}_{0j} \rangle = 0$ if $i \neq j$ and $\langle E_{01} E^{*}_{01} \rangle = \langle E_{02} E^{*}_{02} \rangle$ so that $S_1 = S_2 = S_3 = 0$ and

$$\det \boldsymbol{\rho} = \frac{1}{4} \ .$$

It is customary to define the degree of polarization of partially polarized light by the positive number P defined by

$$\det \boldsymbol{\rho} = \frac{1}{4}(1 - P^2) \ . \tag{50}$$

A.2 Momentum and Angular Momentum Carried by an Electromagnetic Wave.

Let us consider a free electron in the presence of an electromagnetic wave. The electron is set in motion by the incident radiation and emits electromagnetic energy at the rate

$$\frac{dW}{dt} = \frac{8\pi}{3}\left(\frac{e^2}{mc^2}\right)^2 S \ . \tag{51}$$

where S is the intensity of the incident wave as measured by the Poynting vector. The scattered radiation carries no net momentum because the Poynting vector averages to zero over the complete solid angle. Thus, the field surrenders momentum at the rate

$$\frac{dP}{dt} = \frac{1}{c}\frac{dW}{dt} = \frac{8\pi}{3c}\left(\frac{e^2}{mc^2}\right)^2 S \ . \tag{52}$$

This amount of momentum must be acquired by the electron. This result can be deduced directly as follows. Let the incident radiation propagate along $\hat{\boldsymbol{n}}$ and be described by the complex electric and magnetic fields given in Eqs. (1)-(3). We suppose that the electron is located in the plane $\boldsymbol{r}\cdot\hat{\boldsymbol{n}} = 0$. Its equation of motion is

$$m\ddot{\boldsymbol{r}} = -e\boldsymbol{E} - \frac{e}{c}\dot{\boldsymbol{r}} \times \boldsymbol{B} + \frac{2e^2}{3c^3}\dddot{\boldsymbol{r}} \ . \tag{53}$$

Neglecting the effect of the magnetic field in a first approximation (*i.e.*, $v << c$), the complex velocity of the electron is

$$\boldsymbol{v} = \frac{e\boldsymbol{E}_0 e^{-i\omega t}}{im\omega}\left(1 + i\frac{2e^2\omega}{3mc^3}\right)^{-1} . \tag{54}$$

We now recall that the electron is subjected to the Lorentz force $-(e/c)\boldsymbol{v} \times \boldsymbol{B}$. The average value of this force is

$$\begin{aligned}\langle \boldsymbol{F}\rangle &= -\frac{e}{2c}\Re(\boldsymbol{v} \times \boldsymbol{B}^*) = -\frac{e\hat{\boldsymbol{n}}}{2c}\Re(\boldsymbol{v}\cdot\boldsymbol{E}^*)\\ &= \frac{1}{3}\left(\frac{e^2}{mc^2}\right)^2 \frac{\boldsymbol{E}_0\cdot\boldsymbol{E}_0^*\hat{\boldsymbol{n}}}{1+(2e^2\omega/3mc^3)^2} \approx \frac{1}{3}\left(\frac{e^2}{mc^2}\right)^2 \boldsymbol{E}_0\cdot\boldsymbol{E}_0^*\hat{\boldsymbol{n}} \ ,\end{aligned} \tag{55}$$

since $(2e^2\omega/3mc^3) = (2a\omega/3c)$, where a is the classical radius of the electron, is negligible compared to unity for radiation of wavelength $\lambda \gg a$. Using Eq. (36), we find

$$\langle \boldsymbol{F}\rangle = \frac{8\pi}{3c}\left(\frac{e^2}{mc^2}\right)^2 \langle \boldsymbol{S}\rangle = \frac{\sigma}{c}\langle \boldsymbol{S}\rangle \tag{56}$$

where $\boldsymbol{S}$ is the Poynting vector and σ the Thomson scattering cross section for an electron.

We consider now the torque exerted by the electromagnetic wave on the electron. The average torque is

$$\langle \boldsymbol{N}\rangle = -\frac{e}{2}\Re(\boldsymbol{r} \times \boldsymbol{E}^*) \tag{57}$$

where

$$\boldsymbol{r} = \frac{i\boldsymbol{v}}{\omega} \tag{58}$$

is the position of the electron. Then

$$\langle \boldsymbol{N}\rangle = -\frac{e^2}{2m\omega^2}\Re\left(\frac{\boldsymbol{E}_0 \times \boldsymbol{E}_0^*}{1+i(2e^2\omega/3mc^3)}\right) . \tag{59}$$

We now write $\boldsymbol{E}_0$ in the form used in Sec. 2, namely

$$\boldsymbol{E}_0 = a_1 e^{-i\delta_1}\hat{\mathbf{e}}_1 + a_2 e^{-i\delta_2}\hat{\mathbf{e}}_2$$

to obtain

$$\boldsymbol{E}_0 \times \boldsymbol{E}_0^* = 2ia_1a_2\sin\delta\,\hat{\mathbf{e}}_1 \times \hat{\mathbf{e}}_2 \tag{60}$$

where $\delta = \delta_2 - \delta_1$ is the phase difference between the two components of the electric field and $\hat{\mathbf{e}}_1 \times \hat{\mathbf{e}}_2 = \hat{\boldsymbol{n}}$. Thus,

$$\langle \boldsymbol{N}\rangle = -\frac{c\hat{\boldsymbol{n}}}{3\omega}\left(\frac{e^2}{mc^2}\right)^2 2a_1a_2\sin\delta = \frac{c\hat{\boldsymbol{n}}}{3\omega}\left(\frac{e^2}{mc^2}\right)^2 S_3 \tag{61}$$

where S_3 is the Stokes parameter defined in the last Eq. (25). Since for a plane polarized wave $S_3 = 0$, linearly polarized radiation does not transfer a net amount of angular momentum to the electron. For circularly polarized radiation with positive helicity a_1, a_2 are equal and $\delta = -\pi/2$. Thus $S_3 = 2a_1^2$ and $\langle \boldsymbol{S} \rangle = (c\hat{\boldsymbol{n}}/8\pi)S_3$ For circularly polarized radiation with negative helicity $S_3 = -2a_1^2$ but $\boldsymbol{S}$ remains $(c\hat{\boldsymbol{n}}/8\pi)|S_3|$. Thus

$$\langle \boldsymbol{N} \rangle = \pm\frac{\sigma}{\omega}\langle \boldsymbol{S} \rangle \tag{62}$$

where the $\pm$ signs correspond to the two possible helicities of circularly polarized light.

Problems

1. An isolated electron in a magnetic field can be in either of two stationary states denoted here by α and β. The Hamiltonian of the electron is

$$H = 2\mu_B B S_z$$

where the z–direction is taken along that of the magnetic field. $S_z\alpha = \frac{1}{2}\alpha$ and $S_z\beta = -\frac{1}{2}\beta$. Suppose the electron is initially in the state α. What is the half lifetime for decay into state β by spontaneous emission of electromagnetic radiation.

Solution: The transition probability for emission of a photon with polarization $\hat{\mathbf{e}}$ and wave vector $\boldsymbol{k}$ pointing into the element of solid angle $d\Omega$ is

$$dw^{(e)}_{\alpha\to\beta} = \frac{\omega}{2\pi\hbar c^3}|\langle\beta|\hat{\mathbf{e}}\cdot \boldsymbol{J}(\boldsymbol{k})|\alpha\rangle|^2 d\Omega$$

where $\hbar\omega = 2\mu_B B$ and $\boldsymbol{J}(\boldsymbol{k}) = -2ic\mu_B \boldsymbol{k}\times\boldsymbol{S}$. Now $\langle\beta|\hat{\mathbf{e}}\cdot\boldsymbol{J}(\boldsymbol{k})|\alpha\rangle = -ic\mu_B k(\hat{\mathbf{e}}\times\hat{\boldsymbol{k}})\cdot(\hat{\boldsymbol{x}}+i\hat{\boldsymbol{y}})$. For given $\hat{\boldsymbol{k}}$ we select a polarization $\hat{\mathbf{e}}_1$ perpendicular to the plane of $\boldsymbol{B}$ and $\boldsymbol{k}$ and a second $\hat{\mathbf{e}}_2$ in the plane of $\boldsymbol{B}$ and $\boldsymbol{k}$. Adding over the two polarizations (we can select $\hat{\boldsymbol{k}}$ in the $(\boldsymbol{y},\boldsymbol{z})$ plane so that $\hat{\mathbf{e}}_1 = \hat{\boldsymbol{x}}$) we have

$$dw_{\alpha\to\beta} = \frac{\omega\mu_B^2 k^2}{2\pi\hbar c}(1+\cos^2\theta)d\Omega$$

so that

$$w_{\alpha\to\beta} = \frac{8\omega\mu_B^2k^2}{3\hbar c} = \frac{8\mu_B^2}{3\hbar^4c^3}(2\mu_B B)^3 \; .$$

For $B =$ 10,000 gauss, $w \approx 4.4\times 10^{-11}s^{-1}$. Thus, the mean lifetime of the state α is $2.3\times 10^{10}s$ or 729 years.

2. A hydrogen atom in the $2s$ state cannot decay into the $1s$ ground state with emission of a single photon. Calculate the transition probability $2s\to 1s$ with the simultaneous emission of two photons.

Solution: Let $|\nu\rangle$ and $|\nu_0\rangle$ be the initial and final states, respectively. $E_\nu > E_{\nu_0}$. Suppose the radiation field is in its ground state at time $t = 0$ and that the final state contains photons $(\boldsymbol{k}, \hat{e})$ and $(\boldsymbol{k}', \hat{e}')$ such that $\omega + \omega' = (E_\nu - E_{\nu_0})/\hbar = \omega_{\nu\nu_0}$. The matrix elements prior to time integrations are

$$\langle r\nu_0|H'_{II}(t')|r_0\nu\rangle = \frac{2\pi\hbar\hat{e}\cdot\hat{e}'}{V(\omega\omega')^{1/2}}\exp\left[i(\omega+\omega'-\omega_{\nu\nu_0})t'\right]\left\langle\nu_0\left|\sum_i\frac{q_i^2}{m_i}\exp\left(-i(\boldsymbol{k}+\boldsymbol{k}')\cdot\boldsymbol{r}_i\right)\right|\nu\right\rangle \tag{1}$$

and

$$\langle r\nu_0|H'_I(t_2)H'_I(t_1)|r_0\nu\rangle = \sum_{r'\nu'}\langle r\nu_0|H'_I(t_2)H_I(t_1)|r'\nu'\rangle\langle r'\nu'|H_I(t_1)|r_0\nu\rangle \tag{2}$$

where r' can take two possible states in each of which there is a single photon in either of the states $(\boldsymbol{k}, \hat{e})$ or $(\boldsymbol{k}', \hat{e}')$. We have

$$\langle r\nu_0|H'_I(t_2)H'_I(t_1)|r_0\nu\rangle = \sum_{\nu'}\frac{2\pi\hbar}{V(\omega\omega')^{1/2}}\Big\{ \langle\nu_0|\hat{e}'\cdot\boldsymbol{J}(\boldsymbol{k}')|\nu'\rangle\langle\nu'|\hat{e}\cdot\boldsymbol{J}(\boldsymbol{k})|\nu\rangle\exp\left[i(\omega_{\nu_0\nu'}+\omega')t_2+i(\omega_{\nu'\nu}+\omega)t_1\right]$$
$$+\langle\nu_0|\hat{e}\cdot\boldsymbol{J}(\boldsymbol{k})|\nu'\rangle\langle\nu'|\hat{e}'\cdot\boldsymbol{J}(\boldsymbol{k}')|\nu\rangle\exp\left[i(\omega_{\nu_0\nu'}+\omega)t_2+i(\omega_{\nu'\nu}+\omega')t_1\right]\Big\}. \tag{3}$$

We must now perform the integrations over the intermediate times:

$$-\frac{1}{\hbar^2}\int_0^t dt_2\int_0^{t_2}dt_1\langle r\nu_0|H'_I(t_2)H_I(t_1)|r_0\nu\rangle = -\frac{2\pi}{\hbar V(\omega\omega')^{1/2}}\int_0^t dt_2\sum_{\nu'}\Big\{\hat{e}'\cdot\boldsymbol{J}_{\nu_0\nu'}(\boldsymbol{k}')\hat{e}\cdot\boldsymbol{J}_{\nu'\nu}(\boldsymbol{k}) \cdot\frac{e^{i(\omega+\omega'-\omega_{\nu\nu_0})t_2}-e^{i(\omega_{\nu'\nu}+\omega)t_2}}{i(\omega_{\nu'\nu}+\omega)}$$
$$+\hat{e}\cdot\boldsymbol{J}_{\nu_0\nu'}(\boldsymbol{k})\hat{e}'\cdot\boldsymbol{J}_{\nu'\nu}(\boldsymbol{k}')\frac{e^{i(\omega+\omega'-\omega_{\nu\nu_0})t_2}-e^{i(\omega_{\nu'\nu}+\omega')t_2}}{i(\omega_{\nu'\nu}+\omega')}\Big\} \tag{4}$$

where to shorten the notation we wrote

$$\langle\nu|\hat{e}\cdot\boldsymbol{J}(\boldsymbol{k})|\nu'\rangle = \hat{e}\cdot\boldsymbol{J}_{\nu\nu'}(\boldsymbol{k})\ . \tag{5}$$

We define

$$Q_{\nu_0\nu} = \sum_{\nu'}\left(\frac{\hat{e}'\cdot\boldsymbol{J}_{\nu_0\nu'}(\boldsymbol{k}')\hat{e}\cdot\boldsymbol{J}_{\nu'\nu}(\boldsymbol{k})}{\omega_{\nu'\nu}+\omega}+\frac{\hat{e}\cdot\boldsymbol{J}_{\nu_0\nu'}(\boldsymbol{k})\hat{e}'\cdot\boldsymbol{J}_{\nu'\nu}(\boldsymbol{k}')}{\omega_{\nu'\nu}+\omega'}\right)\ . \tag{6}$$

Integration over t_2 yields

$$-\frac{1}{\hbar^2}\int_0^t dt_2 \int_0^{t_2} dt_1 \, \langle r\nu_0 | H'_I(t_2) H'_I(t_1) | r_0 \nu \rangle$$

$$= \frac{2\pi}{\hbar V(\omega\omega')^{1/2}} Q_{\nu_0\nu} \frac{e^{i(\omega+\omega'-\omega_{\nu\nu_0})t}-1}{(\omega+\omega'-\omega_{\nu\nu_0})}$$

$$-\frac{2\pi}{\hbar V(\omega\omega')^{1/2}} \sum_{\nu'} \left\{ \frac{\hat{e}'\cdot \boldsymbol{J}_{\nu_0\nu'}(\boldsymbol{k}')\hat{e}\cdot \boldsymbol{J}_{\nu'\nu}(\boldsymbol{k})}{(\omega_{\nu'\nu}+\omega)^2}\left(e^{i(\omega_{\nu'\nu}+\omega)t}-1\right)\right.$$

$$\left. + \frac{\hat{e}\cdot \boldsymbol{J}_{\nu_0\nu'}(\boldsymbol{k})\hat{e}'\cdot \boldsymbol{J}_{\nu'\nu}(\boldsymbol{k}')}{(\omega_{\nu'\nu}+\omega')^2}\left(e^{i(\omega_{\nu'\nu}+\omega')t}-1\right)\right\} . \qquad (7)$$

We note that the second set of terms can be neglected unless an intermediate state $|\nu'\rangle$ of the system exists for which $\omega_{\nu\nu'}$ equals either ω or ω'. Since no such state exists those terms can be neglected. They would represent actual transitions in first order from $|\nu\rangle$ to $|\nu'\rangle$ with emission of a single photon. Combining the integrated value of (1) and Eq. (7) for two–photon emission we obtain the amplitude

$$\frac{2\pi}{\hbar V(\omega\omega')^{1/2}}\left[Q_{\nu_0\nu} - \hbar\hat{e}\cdot\hat{e}'\left\langle \nu_0 \middle| \sum_i \frac{q_i^2}{m_i}\exp\left(-i(\boldsymbol{k}+\boldsymbol{k}')\cdot \boldsymbol{r}_i\right)\right) \middle| \nu \right\rangle\right] \qquad (8)$$

$$\cdot \frac{e^{i(\omega+\omega'-\omega_{\nu\nu_0})t}-1}{(\omega+\omega'-\omega_{\nu\nu_0})} .$$

If we denote the quantity in square brackets by $\widetilde{Q}_{\nu_0\nu}$, the probability that emission of the photons $(\boldsymbol{k},\hat{e})$ and $(\boldsymbol{k}',\hat{e}')$ has taken place with a transition from $|\nu\rangle$ to $|\nu_0\rangle$ is

$$\frac{4\pi^2}{\hbar^2 V^2(\omega\omega')} \left|\widetilde{Q}_{\nu_0\nu}\right|^2 \frac{\sin^2[(\omega+\omega'-\omega_{\nu\nu_0})t/2]}{[(\omega+\omega'-\omega_{\nu\nu_0})/2]^2} . \qquad (9)$$

We recall that when an actual transition takes place we can replace

$$\frac{\sin^2[(\omega+\omega'-\omega_{\nu\nu_0})t/2]}{[(\omega+\omega'-\omega_{\nu\nu_0})/2]^2} .$$

by $2\pi t\,\delta(\omega_{\nu\nu_0}-\omega-\omega')$. Thus, the probability per unit time for the occurrence of two photon emission is

$$dw_{\nu\to\nu_0} = \frac{8\pi^3}{\hbar^2 V^2}\frac{\left|\widetilde{Q}_{\nu_0\nu}\right|^2}{\omega\omega'}\,\delta(\omega_{\nu\nu_0}-\omega-\omega') . \qquad (10)$$

We must remember that $\widetilde{Q}_{\nu_0\nu}$ depends not only on ω, ω' and the states of the material system but also on $\hat{e}$ and $\hat{e}'$. The total probability per unit time for the transition $|\nu\rangle \to |\nu_0\rangle$ with emission of two photons is now obtained by performing a sum over all $\boldsymbol{k}$ and $\boldsymbol{k}'$ wave vectors and two independent polarizations $\hat{\boldsymbol{e}}$ and $\hat{\boldsymbol{e}}'$ for each of the photons with wave vectors $\boldsymbol{k}$ and $\boldsymbol{k}'$. Now the sum over $\boldsymbol{k}$ can be approximated by integrals over the frequency and the solid angle according to

$$\sum_{\boldsymbol{k}} F(\boldsymbol{k}) = \frac{V}{(2\pi)^3}\int_0^\infty \frac{\omega^2 d\omega}{c^3}\int d\Omega\, F(\boldsymbol{k}) \ .$$

Then, denoting by $\sum_\mu$ $(\mu = 1, 2)$ the summation over the polarizations we have

$$\begin{aligned} w_{\nu\to\nu_0} &= \frac{1}{8\pi^3}\sum_\mu\sum_{\mu'}\frac{1}{\hbar^2 c^6}\int \omega\omega' d\omega d\omega' d\Omega d\Omega' |\widetilde{Q}_{\nu_0\nu}|^2 \delta(\omega_{\nu\nu_0} - \omega - \omega') \\ &= \frac{\hbar}{(2\pi\hbar c^2)^3}\sum_\mu\sum_{\mu'}\int d\Omega d\Omega' |\widetilde{Q}_{\nu_0\nu}|^2 \omega(\omega_{\nu\nu_0} - \omega) d\omega \ . \end{aligned} \tag{11}$$

We consider now the case when the energy difference $E_\nu - E_{\nu_0} = \hbar\omega_{\nu\nu_0}$ is such that the wavelength of the radiation emitted is large compared with the dimensions of the atomic system. Taking the origin of coordinates inside the system we neglect $\boldsymbol{k}\cdot\boldsymbol{r}_i$ so that all exponentials involving $\boldsymbol{k}\cdot\boldsymbol{r}_i$ can be set equal to unity (if the origin of the coordinate system is, say, outside the atom or molecule and $\boldsymbol{R}$ is the center of mass of the molecule, then $\boldsymbol{k}\cdot(\boldsymbol{r}_i - \boldsymbol{R})$ is small compared to unity; all matrix elements are then multiplied by phase factors of absolute value unity and the result is the same as when $\boldsymbol{R} = 0$) For the sake of completeness we mention that we sum the squares of the amplitudes instead of taking the square of the sum since the photons have random phases.

We note that in the electric dipole approximation the term originating from H''_{II} vanishes because of the orthogonality of $|\nu\rangle$ and $|\nu_0\rangle$. The quantity $Q_{\nu_0\nu}$ can be transformed as follows:

We expand $\hat{\boldsymbol{e}}\cdot\boldsymbol{J}(\boldsymbol{k})$ in powers of $\boldsymbol{k}$ in the form

$$\hat{\boldsymbol{e}}\cdot\boldsymbol{J}(\boldsymbol{k}) = \sum_{\ell=0}^{\infty}\hat{\boldsymbol{e}}\cdot\boldsymbol{J}^{(\ell)}k^{(\ell)} \ ,$$

where

$$\boldsymbol{J}^{(0)} = \sum_i \frac{q_i}{m_i}\boldsymbol{p}_i \ ,$$

$$\hat{\boldsymbol{e}}\cdot\boldsymbol{J}^{(1)} = -i\sum_i\left(\frac{q_i}{m_i}\hat{\boldsymbol{e}}\cdot\boldsymbol{p}_i\boldsymbol{r}_i\cdot\hat{\boldsymbol{k}} - c(\hat{\boldsymbol{e}}\times\hat{\boldsymbol{k}})\cdot\boldsymbol{\mu}_i\right) ,$$

and, in general

$$\hat{\boldsymbol{e}}\cdot\boldsymbol{J}^{(\ell)} = \frac{(-i)^\ell}{\ell!}\sum_i\left(\frac{q_i}{m_i}\hat{\boldsymbol{e}}\cdot\boldsymbol{p}_i\left(\boldsymbol{r}_i\cdot\hat{\boldsymbol{k}}\right)^\ell - c\ell\left(\hat{\boldsymbol{e}}\times\hat{\boldsymbol{k}}\right)\cdot\boldsymbol{\mu}_i\left(\boldsymbol{r}_i\cdot\hat{\boldsymbol{k}}\right)^{\ell-1}\right) .$$

We recall that, in the absence of an external magnetic field

$$[\boldsymbol{r}_i, H_0] = \frac{i\hbar \boldsymbol{p}_i}{m_i}$$

so that, defining

$$\boldsymbol{d} = \sum_i q_i \boldsymbol{r}_i$$

we obtain

$$[\boldsymbol{d}, H_0] = i\hbar \sum_i \frac{q_i}{m_i} \boldsymbol{p}_i = i\hbar \boldsymbol{J}^{(0)} . \tag{12}$$

Taking matrix elements of both sides of Eq. (12),

$$\left\langle \nu' | \boldsymbol{J}^{(0)} | \nu \right\rangle = i\omega_{\nu'\nu} \boldsymbol{d}_{\nu'\nu} .$$

We obtain

$$Q_{\nu_0\nu} = -\hat{\boldsymbol{e}}' \cdot \sum_{\nu'} \omega_{\nu_0\nu'} \omega_{\nu'\nu} \left(\frac{\boldsymbol{d}_{\nu_0\nu'} \boldsymbol{d}_{\nu'\nu}}{\omega_{\nu'\nu} + \omega} + \frac{\boldsymbol{d}_{\nu'\nu} \boldsymbol{d}_{\nu_0\nu'}}{\omega_{\nu'\nu} + \omega'} \right) \cdot \hat{\boldsymbol{e}} .$$

Taking into account that $\omega_{\nu\nu_0} = \omega + \omega'$ we find that

$$\frac{\omega_{\nu_0\nu'} \omega_{\nu'\nu}}{\omega_{\nu'\nu} + \omega} = \omega_{\nu_0\nu'} + \omega + \frac{\omega\omega'}{\omega_{\nu'\nu} + \omega}$$

and

$$\begin{aligned} \frac{\omega_{\nu_0\nu'} \omega_{\nu'\nu}}{\omega_{\nu'\nu} + \omega'} &= \omega_{\nu_0\nu'} + \omega' + \frac{\omega\omega'}{\omega_{\nu'\nu} + \omega'} \\ &= \omega_{\nu\nu'} - \omega + \frac{\omega\omega'}{\omega_{\nu'\nu} + \omega'} . \end{aligned}$$

These results allow us to reduce the expression of $Q_{\nu_0\nu}$ to

$$Q_{\nu_0\nu} = -\omega\omega' \hat{\boldsymbol{e}}' \cdot \sum_{\nu'} \left(\frac{\boldsymbol{d}_{\nu_0\nu'} \boldsymbol{d}_{\nu'\nu}}{\omega_{\nu'\nu} + \omega} + \frac{\boldsymbol{d}_{\nu'\nu} \boldsymbol{d}_{\nu_0\nu'}}{\omega_{\nu'\nu} + \omega'} \right) \cdot \hat{\boldsymbol{e}} . \tag{13}$$

The additional terms are identically equal to zero. The first two terms are equal and of opposite signs because the different components of $\boldsymbol{d}$ commute. The commutator of $\hat{\boldsymbol{e}}' \cdot \boldsymbol{J}^{(0)}$ and $\hat{\boldsymbol{e}} \cdot \boldsymbol{d}$ is

$$\left[\hat{\boldsymbol{e}}' \cdot \boldsymbol{J}^{(0)}, \hat{\boldsymbol{e}} \cdot \boldsymbol{d} \right] = -i\hbar \hat{\boldsymbol{e}}' \cdot \hat{\boldsymbol{e}} \sum_i \frac{q_i^2}{m_i}$$

so that

$$\left\langle \nu_0 | [\hat{\boldsymbol{e}}' \cdot \boldsymbol{J}^{(0)}, \hat{\boldsymbol{e}} \cdot \boldsymbol{d}] | \nu \right\rangle = 0$$

because of the orthogonality of $|\nu\rangle$ and $|\nu_0\rangle$. Finally we obtain

$$\begin{aligned} w_{\nu\to\nu_0} = \frac{\hbar}{(2\pi\hbar c^2)^3} \sum_{\mu,\mu'} \int d\omega d\Omega d\Omega' \omega^3 (\omega_{\nu\nu_0} - \omega)^3 \\ \cdot \left| \hat{\mathbf{e}}_{\mu'} \cdot \sum_{\nu'} \left(\frac{\boldsymbol{d}_{\nu_0\nu'}\boldsymbol{d}_{\nu'\nu}}{\omega_{\nu'\nu} + \omega} + \frac{\boldsymbol{d}_{\nu'\nu}\boldsymbol{d}_{\nu_0\nu'}}{\omega_{\nu'\nu_0} - \omega} \right) \cdot \hat{\mathbf{e}}_{\mu} \right|^2 . \end{aligned} \tag{14}$$

Since the $2s$ and $1s$ states are spherically symmetric, the tensor

$$\sum_{\nu'} \left(\frac{\boldsymbol{d}_{\nu_0\nu'}\boldsymbol{d}_{\nu'\nu}}{\omega_{\nu'\nu} + \omega} + \frac{\boldsymbol{d}_{\nu'\nu}\boldsymbol{d}_{\nu_0\nu'}}{\omega_{\nu'\nu_0} - \omega} \right) = D_{\nu_0\nu}\mathbf{1} \tag{15}$$

is a scalar times the unit tensor where we designate any of the diagonal elements of the tensor by $D_{\nu_0\nu}$. Then

$$w_{\nu\to\nu_0} = \frac{\hbar}{(2\pi\hbar c^2)^3} \sum_{\mu,\mu'} \int d\omega d\Omega d\Omega' \omega^3 (\omega_{\nu\nu_0} - \omega)^3 (\hat{\mathbf{e}}_{\mu'} \cdot \hat{\mathbf{e}}_{\mu})^2 |D_{\nu_0\nu}|^2 .$$

The sum over the polarizations is

$$(\hat{\mathbf{e}}_1' \cdot \hat{\mathbf{e}}_1)^2 + (\hat{\mathbf{e}}_1' \cdot \hat{\mathbf{e}}_2)^2 + (\hat{\mathbf{e}}_2' \cdot \hat{\mathbf{e}}_1)^2 + (\hat{\mathbf{e}}_2' \cdot \hat{\mathbf{e}}_2)^2 .$$

Let θ be the angle between $\boldsymbol{k}$ and $\boldsymbol{k}'$ and select $\hat{\mathbf{e}}_1$ and $\hat{\mathbf{e}}_1'$ in the plane of $\boldsymbol{k}$ and $\boldsymbol{k}'$ and $\hat{\mathbf{e}}_2$ and $\hat{\mathbf{e}}_2'$ normal to that plane. Then $(\hat{\mathbf{e}}_1 \cdot \hat{\mathbf{e}}_1')^2 = \cos^2\theta$, $\hat{\mathbf{e}}_1 \cdot \hat{\mathbf{e}}_2' = \hat{\mathbf{e}}_2 \cdot \hat{\mathbf{e}}_1' = 0$, $(\hat{\mathbf{e}}_2 \cdot \hat{\mathbf{e}}_2')^2 = 1$. Thus

$$\sum_{\mu}\sum_{\mu'} (\hat{\mathbf{e}}_{\mu'} \cdot \hat{\mathbf{e}}_{\mu})^2 = 1 + \cos^2\theta .$$

The integrals over the solid angle yield

$$\int d\Omega \int d\Omega' (1 + \cos^2\theta) = \int d\Omega \frac{16\pi}{3} = \frac{64\pi^2}{3} .$$

Then

$$w_{\nu\to\nu_0} = \frac{8}{3\pi\hbar^2 c^6} \int_0^{\omega_{\nu\nu_0}} d\omega \omega^3 (\omega_{\nu\nu_0} - \omega)^3 |D_{\nu_0\nu}|^2 .$$

We now estimate $D(2s \to 1s)$. From Eq. (15)

$$D(2s \to 1s) = e^2 \sum_{\nu} \langle 1s|z|\nu\rangle\langle\nu|z|2s\rangle \left(\frac{1}{\omega_{\nu 2} + \omega} + \frac{1}{\omega_{\nu 1} - \omega} \right) . \tag{16}$$

The states $|\nu\rangle$ having non–zero matrix elements with the $1s$ and $2s$ states are np_0 states and p_0–states in the ionization continuum. We note that for

$|\nu\rangle = |210\rangle$, $\omega_{\nu 2} = 0$ so that we separate that contribution. Thus

$$\begin{aligned} D(2s \to 1s) = {} & e^2 \langle 1s|z|2p_0\rangle\langle 2p_0|z|2s\rangle \left(\frac{1}{\omega} + \frac{1}{\omega_{21}-\omega}\right) \\ & + e^2 {\sum_{\nu}}' \langle 1s|z|\nu\rangle\langle\nu|z|2s\rangle \left(\frac{1}{\omega_{\nu 2}+\omega} + \frac{1}{\omega_{\nu 1}-\omega}\right) . \end{aligned} \tag{17}$$

The sum over ν, indicated by a prime on the summation sign, excludes the term in which $|\nu\rangle = |2p_0\rangle$, *i.e.*, it contains np_0 states with $n \geq 3$ and all p_0 states in the continuous spectrum. Using the wave functions for the H-atom we find

$$\langle 1s|z|2p_0\rangle = a_0\left(2^7\sqrt{2}/3^5\right) ,$$

and

$$\langle 2p_0|z|2s\rangle = -3a_0$$

so that

$$\langle 1s|z|2p_0\rangle\langle 2p_0|z|2s\rangle = -a_0^2\left(2^7\sqrt{2}/3^4\right) = -2.235a_0^2 .$$

Here a_0 is the Bohr radius. An accurate calculation of $D(2s \to 1s)$ is tedious. Therefore we estimate the value of this quantity by keeping the first term. The frequency denominators in the sum in Eq. (17) are larger than those in the first term so that the coefficients

$$\frac{1}{\omega_{\nu 2}+\omega} + \frac{1}{\omega_{\nu 1}-\omega}$$

are positive but less than

$$\frac{1}{\omega} + \frac{1}{\omega_{21}-\omega} .$$

We cannot, however, bracket the result between easily calculated limits because the $\langle 1s|z|\nu\rangle\langle\nu|z|2s\rangle$ do not all have the same sign. The sum

$$\begin{aligned} \sum_{\nu} \langle 1s|z|\nu\rangle\langle\nu|z|2s\rangle &= \langle 1s|z^2|2s\rangle = \tfrac{1}{3}\,\langle 1s|r^2|2s\rangle \\ &= -a_0^2\sqrt{2}(8/9)^3 = -0.993a_0^2 \end{aligned}$$

so that

$${\sum_{\nu}}' \langle 1s|z|\nu\rangle\langle\nu|z|2s\rangle = 10a_0^2\sqrt{2}(2/3)^6 = 1.242a_0^2 .$$

We conclude that a suitable estimate of $w(2s \to 1s)$ is

$$\begin{aligned} w_{\nu\to\nu_0} &= \left(8/3\pi\hbar^2c^6\right)(ea_0)^4\left(2^{15}/3^8\right)\int_0^{\omega_{21}} d\omega\omega(\omega_{21}-\omega)\omega_{21}^2 \\ &= (4/3^5\pi)\alpha^8\left(\frac{mc^2}{\hbar}\right) = 33s^{-1} , \end{aligned}$$

where α is the fine structure constant. Thus

$$\tau = w^{-1}(2s \to 1s) = 0.03s \ .$$

A more accurate estimate has been carried out by Breit and Teller.[21] Their calculation gives $\tau = (1/7)s$. A second possibility to be considered is a transition from the $2s$ state to $2p$ and from $2p$ to $1s$. This mechanism is not allowed because $2s$ and $2p$ are degenerate states. They are even degenerate within the frame of the Dirac relativistic theory. The $2S_{1/2}$ state of hydrogen is degenerate with the $2P_{1/2}$ and the $2P_{3/2}$ levels lies above it. However, the degeneracy of $2P_{1/2}$ and $2S_{1/2}$ is lifted by the interaction with the electromagnetic field and $2P_{1/2}$ lies below $2S_{1/2}$. The energy separation is given by $\nu = 1058 \times 10^6 s^{-1}$ or $\sim 4.4 \times 10^{-6}$ eV. Thus, even though the transition from $2P_{1/2}$ to $1S_{1/2}$ has a lifetime of about $1.6 \times 10^{-9}s$, the transition from $2S_{1/2}$ to $2P_{1/2}$ is so slow that the two photon process is considerably more likely. Collisions with other atoms allows $2S_{1/2}$ to $1S_{1/2}$ transitions[22] to proceed much faster than by two–photon emission but the gas must be dense enough to permit frequent collisions.

References and Footnotes

1. S. Bhagavantam, *The Scattering of Light and the Raman Effect* (Andhara University, Waltair, India, 1940).

2. Leonardo da Vinci, *The Notebooks of Leonardo da Vinci*, rendered into English by E. Mac Curdy (G. Braziller, New York, 1956) p. 399.

3. C. V. Raman, Ind. Journal of Physics **2**, 387 (1928).

4. A. Smekal, Naturwissensch. **11**, 873 (1923).

5. H. A. Kramers and W. Heisenberg, Zeit. f. Physik **31**, 681 (1925).

6. G. Landsberg and L. Mandel'shtam, Naturwissensch. **16**, 57 (1928); *ibid.*, **16**, 772 (1928).

7. A discussion of the Clebsch-Gordan oefficients and a proof of the Wigner-Eckart theorem is given in most texts on quantum mechanics. See, for example, the notes on *"Symmetry in Physics"* by Sergio Rodriguez in the series Appunti published by Scuola Normale Superiore, Pisa, Italy.

8. For a more detailed presentation of the theory of group representations see Ref. 7.

9. The chemical notation for the irreducible representations of finite groups can be found in E.B. Wilson, Jr., J. C. Decius, and P. C. Cross, *"Molecular Vibrations, The Theory of Infrared and Raman Vibration Spectra"*, (McGraw-Hill, New York, 1955).

10. In addition to the nomenclature in Ref. 9, we make use of the notation of G. F. Koster, J. O. Dimmock, R. G. Wheeler, and H. Statz, *"Properties of the Thirty Two Point Groups"*, (MIT Press, Cambridge, MA, 1963).

11. The reduction of the representations Γ_{int} and $[V \times V]$ generated by the components of a symmetric rank tensor such as $\boldsymbol{\alpha}$ are

$$\Gamma_{int} = A_{1g} + E_g + T_{2g} + 2T_{1u} + T_{2u} = \Gamma_1^+ + \Gamma_3^+ + \Gamma_5^+ + 2\Gamma_4^- + 2\Gamma_5^-$$

and

$$[V \times V] = [\boldsymbol{\alpha}] = A_{1g} + E_g + T_{2g} = \Gamma_1^+ + \Gamma_3^+ + \Gamma_5^+ \ ,$$

respectively. Here we give both the chemical notation for the irreducible representation and that of Koster *et al.*. (see Refs. 9 and 10.)

12. V. J. Tekippe, A. K. Ramdas and S. Rodriguez, Phys. Rev. B **8**, 706 (1973).

13. M. H. Grimsditch, A. K. Ramdas, S. Rodriguez and V. J. Tekippe, Phys. Rev. B **15**, 5869 (1977).

14. H. Kim, R. Vogelgesang, A. K. Ramdas, S. Rodriguez, M. Grimsditch, and T. R. Anthony, Phys. Rev. B **57**, 15315 (1998).

15. H. Kim, A. K. Ramdas, S. Rodriguez, M. Grimsditch, and T. R. Anthony, Phys. Rev. Lett. **83**, 3254 (1999).

16. H. Kim, A. K. Ramdas, S. Rodriguez, M. Grimsditch, and T. R. Anthony, Phys. Rev. Lett. **83**, 4140 (1999).

17. H. Kim, Z. Barticevic, A. K. Ramdas, S. Rodriguez, M. Grimsditch, and T. R. Anthony, Phys. Rev. B **62**, 8038 (2000).

18. F. C. von der Lage and H. A. Bethe, Phys. Rev. **71**, 612 (1947).

19. G. Frobenius and I. Schur, Sitz. Berichte Preuss. Akad. Wiss. Phys-math Classe v. 8 (Berlin, Feb. 1906) pp. 186-208; see also V. Heine, *Group Theory in Quantum Mechanics* (Pergamon, New York, 1960) pp. 169-174.

20. J. M. Luttinger, Phys. Rev. **102**, 1030 (1956).

21. G. Breit and E. Teller, Astrophysical Journal **91**, 215 (1940).

22. E. M. Purcell, Astrophysical Journal **116**, 457 (1952).

Elenco dei volumi della collana
"Appunti"
pubblicati dall'Anno Accademico 1994/95

GIUSEPPE BERTIN (a cura di), *Seminario di Astrofisica,* 1995.

EDOARDO VESENTINI, *Introduction to continuous semigroups,* 1996.

LUIGI AMBROSIO, *Corso introduttivo alla Teoria Geometrica della Misura ed alle Superfici Minime,* 1997 (ristampa).

CARLO PETRONIO, *A Theorem of Eliashberg and Thurston on Foliations and Contact Structures,* 1997.

MARIO TOSI, *Introduction to Statistical Mechanics and Thermodynamics,* 1997.

MARIO TOSI, *Introduction to the Theory of Many-Body Systems,* 1997.

PAOLO ALUFFI (a cura di), *Quantum cohomology at the Mittag-Leffler Institute,* 1997.

GILBERTO BINI, CORRADO DE CONCINI, MARZIA POLITO, CLAUDIO PROCESI, *On the Work of Givental Relative to Mirror Symmetry,* 1998

GIUSEPPE DA PRATO, *Introduction to differential stochastic equations,* 1998 (seconda edizione)

HERBERT CLEMENS, *Introduction to Hodge Theory,* 1998

HUYÊN PHAM, *Imperfections de Marchés et Méthodes d'Evaluation et Couverture d'Options,* 1998

MARCO MANETTI, *Corso introduttivo alla Geometria Algebrica,* 1998

AA.VV., *Seminari di Geometria Algebrica 1998-1999,* 1999

ALESSANDRA LUNARDI, *Interpolation Theory,* 1999

RENATA SCOGNAMILLO, *Rappresentazioni dei gruppi finiti e loro caratteri,* 1999

SERGIO RODRIGUEZ, *Symmetry in Physics,* 1999

F. STROCCHI, *Symmetry Breaking in Classical Systems and Nonlinear Functional* Analysis, 1999

LUIGI AMBROSIO, PAOLO TILLI, *Selected Topics on "Analysis in Metric Spaces",* 2000

SERGEI V. BULANOV, *Lectures on Nonlinear Physics,* 2000

LUCA CIOTTI, *Lectures Notes on Stellar Dynamics,* 2000

SERGIO RODRIGUEZ, *The scattering of light by matter,* 2001

"CompoMat" Loc. Braccone, 02040 Configni (RI), Italy
Finito di stampare per conto della "CompoMat" dalla Nuova Grafica 86 nel febbraio 2002

Sergio Rodriguez
Department of Physics
Purdue University
West Lafayette, IN 47907-1396, U.S.A.

The scattering of light by matter